General Purpose Technologies and Economic Growth

MECHANICS' INSTITUTE LIBRARY
57 Post Street
San Francisco, CA 94104
(415) 393-0101

General Purpose Technologies and Economic Growth

edited by Elhanan Helpman

The MIT Press
Cambridge, Massachusetts
London, England

© 1998 Massachusetts Institute of Technology

All rights reserved. No part of this book may be reproduced in any form by any electronic or mechanical means (including photocopying, recording, or information storage and retrieval) without permission in writing from the publisher.

This book was set in Palatino on the Monotype "Prism Plus" PostScript Imagesetter by Asco Trade Typesetting Ltd., Hong Kong.

Printed and bound in the United States of America.

Library of Congress Cataloging-in-Publication Data

General purpose technologies and economic growth / edited by Elhanan Helpman.
   p. cm.
Includes bibliographical references and index.
ISBN 0-262-08263-2 (hc. : alk. paper)
  1. Technological innovations—Economic aspects. 2. Economic development.
I. Helpman, Elhanan.
HC79.T4G464   1998
338.9—dc21   98-11842
                                                                         CIP

to Fraser Mustard, a man of vision

Contents

Preface   ix
Contributors   xi

1 **Introduction**   1
   Elhanan Helpman

2 **What Requires Explanation?**   15
   Richard G. Lipsey, Cliff Bekar, and Kenneth Carlaw

3 **A Time to Sow and a Time to Reap: Growth Based on General Purpose Technologies**   55
   Elhanan Helpman and Manuel Trajtenberg

4 **Diffusion of General Purpose Technologies**   85
   Elhanan Helpman and Manuel Trajtenberg

5 **On the Macroeconomic Effects of Major Technological Change**   121
   Philippe Aghion and Peter Howitt

6 **The Internet as a GPT: Factor Market Implications**   145
   Richard G. Harris

7 **Chemical Engineering as a General Purpose Technology**   167
   Nathan Rosenberg

8 The Consequences of Changes in GPTs 193
Richard G. Lipsey, Cliff Bekar, and Kenneth Carlaw

9 Measurement, Obsolescence, and General Purpose Technologies 219
Peter Howitt

10 The Division of Inventive Labor and the Extent of the Market 253
Timothy Bresnahan and Alfonso Gambardella

11 Wages, Skills, and Technology in the United States and Canada 283
Kevin M. Murphy, W. Craig Riddell, and Paul M. Romer

Index 311

Preface

This book consists of a collection of original essays on a subject at the heart of the study of modern economic growth. Some of the chapters were written before the idea of the book was conceived, others especially for the book. All chapters except one were refereed and revised. The one chapter not refereed anonymously was subjected to detailed criticism by a number of the participants in this endeavor, including the editor. What appears here is a revised version that takes account of those comments.

Most authors of these chapters are associated with the Economic Growth and Policy Program of the Canadian Institute for Advanced Research (CIAR). CIAR is a unique institution that enables researchers from different parts of the world to interact with each other. It is a true university without walls. Fraser Mustard, the founder and first president of CIAR, and Stefan Dupré, its current president, made every effort to encourage this line of research. I thank them both. This is also an opportunity to express our thanks to the Royal Bank of Canada, Canadian Pacific Ltd., Petro-Canada, Alcan Aluminum Ltd., the Bank of Montreal, and the John Dobson Foundation (Montreal) for their financial support. And finally, many thanks to vice president Kathryn Hough, Joanna Lipsey, and Sue Schenk for making the Economic Growth and Policy Program run as smoothly as it does.

Contributors

Philippe Aghion
University College London

Cliff Bekar
Simon Fraser University

Timothy Bresnahan
Stanford University

Kenneth Carlaw
Simon Fraser University

Alfonso Gambardella
University of Urbino

Richard Harris
Simon Fraser University and CIAR

Elhanan Helpman
Harvard University
Tel Aviv University and CIAR

Peter Howitt
Ohio State University and CIAR

Richard G. Lipsey
Simon Fraser University and CIAR

Kevin M. Murphy
University of Chicago and CIAR

W. Craig Riddell
University of British Columbia and CIAR

Paul M. Romer
Stanford University and CIAR

Nathan Rosenberg
Stanford University and CIAR

Manuel Trajtenberg
Tel Aviv University and CIAR

General Purpose Technologies and Economic Growth

1 Introduction

Economic growth is driven by many factors: economic, political, and cultural. While economists have traditionally emphasized the accumulation of conventional inputs (e.g., labor and capital) as the primary force behind output expansion, more recently greater attention has been paid to political and technological factors. This book is concerned with the latter.

How important are technological factors in explaining modern economic growth? Economic historians have placed great weight on technology as a force of change.[1] Macroeconomists, on the other hand, used to downplay its role, mostly due to the inability to analyze forces that shape technological change. In recent years, however, this attitude has changed. Following the work of Romer (1990), Grossman and Helpman (1991), and Aghion and Howitt (1992), many macroeconomic studies now place technological progress at the center of the growth process. This change of heart has been triggered by theoretical developments that allow microeconomic aspects of the innovation process to be linked with macroeconomic outcomes.

Importantly, there have always existed economists who consider technological change to be at the heart of the growth process. A salutary example is Simon Kuznets, who wrote more than thirty years ago:

... we may say that certainly since the second half of the nineteenth century, the major source of economic growth in the developed countries has been science-based technology—in the electrical, internal combustion, electronic, nuclear, and biological fields, among others. (Kuznets 1966, p. 10)

---

1. See, for example, Landes (1969), Rosenberg (1982), and Mokyr (1990).

For this reason it is more accurate to describe the current preoccupation with technology as a revival of an old tradition rather than the development of a new one.[2]

How should economists think about technological progress? The prevalent approach has been to view it as an incremental process that improves the efficiency of resource deployment. It may not be uniform across sectors or time, but the aggregate effects are relatively "smooth."

The incremental nature of technological progress has been well documented by economic historians.[3] But so were major inventions that had far-reaching and prolonged implications, such as the steam engine, electricity, and the computer.[4] Is is useful to distinguish between incremental and drastic innovations? There is hardly a debate about incremental innovations. Small improvements take place in the regular course of business, both serendipitous and intentional. In addition many incremental innovations follow drastic innovations. The introduction of the first steam engine, for example, triggered a sequence of secondary innovations that were designed to improve its operation, and the same is true about electricity and the computer.

Economists have payed less attention to the role of drastic innovations. A drastic innovation introduces a discontinuity, in the sense that it leads to the replacement of an old technology that played a significant role in an industry with new methods of production. Or it replaces an old material that performed an array of designated functions (e.g., rubber) with a new one (e.g., plastics). A discontinuity in this sense does not imply, however, a necessary discontinuity in the observed pattern of resource allocation or the evolution of output, since the introduction of a superior technology can be gradual, starting with a negligible absorption of resources and followed by continuous expansion over time.

But even if a drastic innovation penetrates an economy gradually and the pattern of resource allocation and the level of output change only slowly, it does not mean that it is not helpful to think about drastic innovations. Since by thinking only in terms of incremental innovations we miss the true cause of incremental innovations that are triggered by a

---

2. Even during the years when macroeconomists shied away from technological progress there were other economists, such as Christopher Freeman, Zvi Griliches, Edwin Mansfield, and Nathan Rosenberg, who spent considerable time studying economic aspects of technology.
3. See Rosenberg (1982).
4. See von Tunzelmann (1978) on the steam engine, Du Boff (1967) on electricity, and David (1991) on the parallels between electricity and the computer.

drastic innovation, and we may wrongly attribute all changes in resource allocation and output to the *induced* secondary innovations. It follows that in order to properly understand the relationship between cause and consequence, it is useful to distinguish between drastic and incremental innovations. A drastic (or major) innovation often sets the stage for a series of incremental innovations.

Second, it is possible that forces that drive incremental innovations are different from those that drive drastic innovations. For example, incremental innovations are more susceptible to standard profitability calculations, even when they involve externalities and are subject to risk, simply because markets can evaluate their profitability. In comparison, drastic innovations face much larger uncertainties, producing risks that are much harder to evaluate and therefore harder to insure.[5] As a result drastic innovators can engage in little risk-sharing and have to bear most of the risk themselves.

Finally some drastic innovations may produce discontinuities, if not at the economywide level then at the industry level. In interpreting data, we will not search for such discontinuities if we are preconditioned to think only about incremental innovations. As a result we may misinterpret the evidence.

This book is concerned with a certain type of drastic innovations, termed *general purpose technologies* (GPTs). A drastic innovation qualifies as a GPT if it has the potential for pervasive use in a wide range of sectors in ways that drastically change their modes of operation. To quote from Bresnahan and Trajtenberg (1995), who coined the term GPT and provided a highly original discussion of its usefulness,

> Most GPTs play the role of "enabling technologies," opening up new opportunities rather than offering complete, final solutions. For example, the productivity gains associated with the introduction of electric motors in manufacturing were not limited to a reduction in energy costs. The new energy sources fostered the more efficient design of factories, taking advantage of the newfound flexibility of electric power. Similarly, the users of micro-electronics benefit from the surging power of silicon by wrapping around the integrated circuits their own technical advances. This phenomenon involves what we call "innovational complementarities" (IC), that is, the productivity of R&D in a downstream sector increases as a consequence of innovation in the GPT technology. These complementarities magnify the effects of innovation in the GPT, and help propagate them throughout the economy.

---

5. See Rosenberg (1996).

This description makes clear two important features of drastic technological innovations that qualify as GPTs: generality of purpose and innovational complementaries. When these effects are particularly strong, as, for example, in the case of electricity, they lead to monumental changes in economic organizations. Sometimes they also affect the organization of society through working hours, constraints on family life, social stratification, and the like.

As Bresnahan and Trajtenberg (1995) noted, GPTs introduce two types of externalities: one between the GPT and the application sectors; another across the application sectors. The former stems from the difficulties that a GPT inventor may have in appropriating the fruits of her invention. When institutional conditions prevent full appropriation, the GPT is effectively underpriced and therefore undersupplied. The latter stems from the fact that, since the application sectors are not coordinated, each one conditions its expansion on the available general purpose technology. But if they coordinated a joint expansion, they would raise the profitability of the GPT and encourage its improvement. A better GPT benefits them all.

The following chapters do not provide a single view of GPTs. They were written by individual authors who share the belief that GPTs are important technologies and that identifying their characteristics and economic implications is most desirable. But this belief has not lead everyone to have precisely the same view of the nature of GPTs nor of their implications. For these reasons various controversies can be detected in the pages of the book. By bringing these essays together, however, we hope to encourage further exploration of this subject. For example, at the moment an agreed-upon theoretical framework for dealing with the complex dynamics that are associated with general purpose technologies does not exist. Some of the chapters explore alternative formulations and their implications. This sort of analysis should improve our understanding of GPT-driven economic growth, on the one hand, and lay the ground for further improvements of available frameworks, on the other. Taken together these essays provide some essential background on GPTs and take initial steps in the implementation of a research agenda that could greatly benefit from a broad-based network of researchers.

Chapter 2 contains a detailed discussion of the nature of general purpose technologies that goes beyond the original Bresnahan and Trajtenberg (1995) paper. After a brief review of some of the major theoretical contributions to the subject, Lipsey, Bekar, and Carlaw discuss GPTs throughout history. They describe the nature of information and communication technologies (e.g., printing), materials, power delivery systems (e.g., steam) and transportation (railways and motor vehicles). Drawing on theoretical and historical insights they propose a characterization of GPTs in terms of four essential features: (1) much scope for improvement initially, (2) many varied uses, (3) applicability across large parts of the economy, and (4) strong complementarities with other technologies. To check whether this characterization is sensible, Lipsey, Bekar, and Carlaw proceed to examine various technologies (e.g., lasers) to see which are included in this definition and which are excluded. Importantly, they include organizational technologies as part of the technological landscape, as I believe we should. Given the nature of GPTs there is no reason not to think about the factory system, for example, as a general purpose technology. Lipsey, Bekar, and Carlaw make a strong case for the usefulness of the GPT concept.

In chapter 3 Helpman and Trajtenberg develop a simple theoretical model in order to study the macroeconomic consequences of GPTs. They use a variant of the now standard model in which monopolistically competitive companies develop intermediate inputs and sell them to manufacturers of a final consumer good. They assume away, however, endogenous growth mechanisms, such as sustained learning in the R&D sector, so that long-run growth does not occur, unless new general purpose technologies continue to arrive. To capture the notion of innovational complementarities, they assume that each GPT requires its own intermediate inputs. And to capture the notion of generality of purpose, they first analyze an economy with one final output sector. Their GPTs arrive regularly at fixed intervals of time. This framework generates long-run cycles. In the first phase of each cycle, which is characterized by a period of secondary innovations during which the new technology has not yet been implemented in manufacturing, real output declines. It rises fast in the second phase, which begins with the adoption of the new technology by manufacturers of the final product. Real wages are constant in the first phase and rise in the second. Although the real output decline is an artifact of some technical features of the model, the productivity showdown is an inherent part of this phase. The model thereby captures some features of the cycle associated with the introduction of electricity and possibly computers.

An extension of the basic model introduces a continuum of final-good sectors with different degrees of productivity gains from GPTs and distinguishes between unskilled and skilled labor. Skilled workers are used intensively in research and development; unskilled in manufacturing. In this formulation the intermediate inputs are common to all sectors. As a result a GPT spreads gradually across the economy, first to sectors that stand to gain most from its adoption and last to sectors that stand to gain least. In the process there is a cycle in the relative wage. Skilled workers do relatively better in the first phase, when secondary innovations are developed but no sector has yet adopted the new technology, and less so in the second phase, when more and more sectors switch to the new technology. Now there is a productivity slowdown in the first phase but no decline of real GDP.

Economies of this type suffer from a serious coordination problem. Developers of intermediate inputs for a new GPT invest in R&D in expectation that the new technology will be adopted by application sectors. But application sectors switch to the new technology only when there are enough intermediate inputs to support it. Therefore the profitability of R&D depends on how many other intermediate inputs will be developed.

Companies that specialize in intermediate inputs do not coordinate their development efforts. As a result an equilibrium in which no secondary innovations take place always exists. In this equilibrium each product developer is pessimistic, expecting none of the others to develop new intermediates. Therefore she expects the application sectors not to adopt the new technology, for lack of enough intermediates to make it worthwhile. Under these circumstances it is not possible to make a profit on the development of a new intermediate input. Such expectations prove to be self-fulfilling, and the new technology is never adopted. Clearly this is an extreme example of the type of coordination failures discussed by Bresnahan and Trajtenberg (1995). Equilibria of this type, in which GPTs are bypassed, coexist with equilibria in which they are implemented.

Some implications of these coordination problems at the sectoral level are explored in chapter 4. The basic model is the same as in the previous chapter, except that each sector requires its own intermediate inputs for every GPT. As a result there is no natural order in which the application sectors adopt a new general purpose technology. Instead, multiple self-fulfilling expectations equilibria exist that differ in their order of adoption, and they differ in their welfare implications. But there is no apparent market mechanism that can select an equilibrium with the most efficient order of adoption.

For every feasible order of adoption, a cycle is associated with every industry. In the first phase of the cycle sector-specific intermediates are developed while manufacturers of the final good continue to use the old technology. In the second phase those producers use the new technology. But after all sectors go through these cycles, a second wave of secondary innovations occurs in which all application sectors participate.

Four parameters characterize a sector: (1) the productivity advantage of the new GPT, (2) the historically determined stock of past investments in the form of available intermediates, (3) the level of demand, and (4) the productivity of resources in secondary innovation. The combination of all four determines the desired order of adoption. As a reality check on the working of this model, Helpman and Trajtenberg examine the early diffusion of the transistor. They find that the early adopters (hearing aids and computers) had favorable conditions in terms of all four parameters. As for the laggards (telecommunications and automobiles), they appear to have had large, historically determined stocks of past investments.

A modified version of the Helpman-Trajtenberg model is developed in chapter 5 by Aghion and Howitt. To begin with, they show that similar dynamics emerge when the intermediate inputs are vertically rather than horizontally differentiated. For this purpose they adapt the original Aghion and Howitt (1992) model to general purpose technologies. As in chapters 3 and 4 a GPT arrives first. Once available, companies invest in the development of a suitable intermediate input. This produces the first phase of a GPT-related cycle when labor is drawn out of manufacturing and into R&D. The fall in output is attributed to the fact that research and development is undervalued in the national accounts. Phase one continues as long as the suitable input has not been invented, which is stochastic, with the average length declining as more workers are employed in R&D. Once a suitable input has been invented, the economy moves to phase two of the cycle, in which the new technology is implemented and all manufacturers of final output use the new intermediate input together with the new GPT.

Aghion and Howitt argue that this theoretical structure suffers from two empirical shortcomings. First, the size of the slump produced in the first phase is too mild, since it is determined by employment in research and development which is empirically rather small. Second, the slowdown appears too early as compared with the historical evidence. For these reasons they propose a modification. Instead of having two stages in the adoption of a new GPT, a first stage in which new intermediates are developed and a second stage in which they are used in manufacturing, Aghion and Howitt propose an additional stage. Before an intermediate input can be developed, they argue, companies need to discover a "template" on which research can be based. This "template" is specific to each

GPT. In addition "templates" and intermediate inputs are sector specific (just like the intermediate inputs in chapter 4). Importantly, technological spillovers occur across sectors. Costs of "template" development for a particular industry decline the more industries have already developed "templates" for the new GPT. Externalities of this sort emanate from learning; the experience of others in coping with the new technology is helpful to our own efforts. This feature in combination with the intermediate input development stage produce a delayed slump in output that is approached only gradually. But it is also possible that no slump occurs. Instead a slowdown in productivity growth is followed by acceleration.

To account for the size of the slump, Aghion and Howitt propose considering some additional features. First (as in chapter 3), a distinction between skilled and unskilled workers, with the former used more intensively in the implementation of the new technology. This helps to account for a productivity slowdown and also for movements in the relative wage of the two groups. Second, costly search in the labor market that produces frictional unemployment leads to a temporary rise in unemployment with the arrival of a new GPT. Finally, capital obsolescence caused by the wave of secondary innovations also contributes to a decline in output. Whether these features can account for observed patterns of GPT-related output movements is an empirical question that has not been resolved.

In chapter 6 Harris takes up international trade-related issues, and in particular the effects of GPTs on factor markets. His focus is on GPTs that enhance international trade in services, the Internet being his prime motivation. Most manufacturers use a variety of services as intermediate inputs, such as accounting, inventory management, design, and programming. Such services had typically to be provided in-house, or at least in proximity to a company's major operations. By drastically reducing the costs of acquiring such services from distant suppliers, the Internet has virtually eliminated the proximity requirement. Now many of these services can be easily purchased from distant suppliers, including those from other countries, at little extra cost.

Constructing a standard model with two sectors, one producing with constant returns to scale and the other with intermediate inputs that are produced under increasing returns, Harris explores the effects of a move from a world in which services are not traded internationally to one in which they are. A typical country (or region) employs skilled and unskilled workers in manufacturing. Intermediate inputs are skill intensive. Under these circumstances, absent international trade in services, countries with more skilled workers have more specialized services available for manufacturing. As a result the wage rate of the skilled is higher and the wage rate of the unskilled is lower. It follows that the skill premium is larger in countries with relatively more skilled workers.

In comes the Internet. It allows the trading of services at low cost, and it creates familiar pricing problems. Assuming average cost pricing of network services, Harris explores the effects of trade in intermediate inputs (services) on factor markets. The Internet leads to the supply of more specialized intermediate inputs and raises the demand for skill. In response the wage rate of skilled workers rises and the wage rate of the unskilled falls. The skill premium rises. In addition total factor productivity rises, and each economy produces more output as measured by GDP.

Using these insights, Harris explores a set of additional issues, such as regional differences in the post Internet era and incentives to adopt the network.

The engineering professions are relatively new. Rosenberg argues in chapter 7 that they need to be thought as the repositories of technological knowledge, and their practitioners as the primary agents of technological change in their respective industries. In this context he argues that chemical engineering can be usefully thought of as a general purpose technology.

According to Rosenberg chemical engineering did not have its origins in science, although at a later stage in its development it began to draw on science. Nevertheless, it consists of a body of useful knowledge concerning manufacturing processes. As such it helps to guide the design and operation of chemical plants.

A central concept of the discipline, the "unit operations" (e.g., mixing, heating, and crystallizing), was articulated by Arthur D. Little in 1915. Its centrality derives from the fact that there exists a relatively small number of unit operations and that any chemical process can be decomposed into a series of unit operations. This characterization provides it with a key attribute that makes chemical engineering a GPT. As such it has provided valuable services to a large number of industries, which would not necessarily be identified with the chemical sector. They include rubber, leather, sugar refining, and glass.

Rosenberg reviews the history of the discipline, the development and codification of its concepts, and the penetration of these concepts to application sectors. Prominent among the application sectors was petroleum refining, triggered by the fast-growing automobile industry. Along with the development of research in polymer chemistry, this experience fostered the development of a range of new products from petroleum feedstocks, such as synthetic fiber and synthetic rubber. The "unit operations" concept was indispensable to the secondary innovations that followed the development of the discipline.

A detailed description of the channels through which GPTs affect an economy is provided in chapter 8 by Lipsey, Bekar, and Carlaw. They propose to replace the black box of the "production function," which is widely used in economic analysis, with an explicit treatment of the interactions between technology, facilitating structures, and public policy. Recognizing such interactions, they argue, is important for a proper understanding of general purpose technologies.

To make their case, Lipsey, Baker, and Carlaw trace the effects of a new GPT on other technologies in use, those that it is bound to replace as well as technologies of a complementary nature. In addition they trace its effects on the facilitating structure, which they view as "the embodiment of technological and policy knowledge," and on public policy. The multitude of channels of influence that emerges from this discussion leaves plenty of room for the empirical researcher, who now needs to provide an evaluation of which channels have particularly large quantitative effects.

Next Lipsey, Bekar, and Carlaw examine the effects of GPTs on performance, first in labor markets and second in terms of GDP. Here short-run effects are distinguished from long-run effects, and transitional dynamics receive due attention. The chapter closes with a short list of challenges for future research.

An arrival of a general purpose technology triggers a chain of activities. Some of them are specific to GPTs, as described in previous chapters. Others apply to additional forms of technological change. In particular the arrival of a new GPT accelerates the pace of technological progress. This acceleration of pace, although not unique to GPTs, has important consequences.

Howitt tackles in chapter 9 an especially important implication of faster technological progress, namely the rise in obsolescence of capital. As he points out, it applies to machines and equipment as well as to human capital. Such obsolescence leads to a slowdown in productivity growth. But measurement problems abound during turbulent times of major technological change, and the mismeasurement of obsolescence is just one of them.

Three measurement problems are reviewed by Howitt. First, many of the knowledge-creating activities are not registered as part of GDP under conventional national accounting. A company that employs workers in research and development for internal use, for example, registers a fall in profits equal in value to the wages of these workers, who produce no tangible output. Some of the benefits of this R&D activity will show up in the future, but these benefits are not capitalized under conventional accounting practices. Second, the contribution of new or improved products to output is typically underestimated. This is a well-known problem that becomes more severe under rapid technological change. Finally, the arrival of new technologies makes machines and equipment that were designed for the old one obsolete, and it reduces the value of skills that workers acquired for the old technology. This destruction of productive assets is not accounted for the most part. And it gains importance when the pace of technological progress accelerates.

To assess the magnitude of the possible bias introduced by the first and last problem, Howitt develops a simple analytical model and calibrates it to U.S. data. Simulations of this model suggest that the obsolescence problem is quantitatively more important than the underestimation of the value of R&D activities and that the transitional dynamics last for many years. When the pace of technological progress rises by 50 percent, measured GDP declines by 8.5 percent after 22 years (the lowest point). If employment in R&D were valued at cost, however, the decline in output would have reached the lowest point after 15 years, with a 6.6 percent shortfall. The rest is driven by obsolescence. These phenomena seem to be quite important, and they deserve much more attention.

In chapter 10 Bresnahan and Gambardella examine the effects of GPTs on incentives for vertical integration. They relate this issue to the degree of specialization of labor. As a result they also shed new light on Adam Smith's famous arguments about the relationship between the degree of specialization and the extent of the market.

For Bresnahan and Gambardella there are two measures of the extent of the market: the output level of an industry and the range of relevant industries. Most discussions use the first measure, and in particular in industrial organization. They therefore conclude the opposite from Smith; namely, as the extent of the market expands (output rises), the incentives for vertical integration increase and the degree of specialization declines. The authors show, however, that the opposite is true when the extent of the market is defined in terms of the range of industries. That is to say, as the range of industries that define the extent of the market increases, the incentive for vertical integration declines and the degree of specialization increases.

Important in this analysis is the characterization of the technology, which is a GPT. A new GPT is suitable for a range of industries. Manufacturers can use it with little adaptation, or resources can be invested to adapt it to each product line (secondary innovations). In the former case little specialization occurs; in the latter, significant specialization. For cost-minimizing industries, expanding the output of each product line encourages the use of a GPT with little adaptation. In contrast, expansion of the range of industries for which the GPT is useful encourages secondary innovations and specialization. The specialization trend is stronger the lower the adaptation costs.

To give concrete content to these theoretical distinctions, the authors re-examine the example of railroads, which Stigler used in his seminal work on this subject. They interpret the evolution of the railroad in view of the model's parameters. In addition they argue that chemical engineering, based on the concept of the "unit operations" (discussed in chapter 7) can also be interpreted in a similar light.

In further theoretical development Bresnahan and Gambardella consider the supply of GPTs. When designing a GPT, the supplier can make it more or less adaptable to a wide variety of users. More adaptability involves, however, higher development costs. Once a GPT is available, application sectors can choose to use it with or without secondary innovations. A fraction of the heterogeneous application sectors, which differ in demand levels, chooses the adaptation option. Now it is possible to study the effects of the two measures of the extent of the market on the size of the GPT sector. Its size rises with the expansion of the range of potential application sectors but declines with a rise in the demand level of each application sector. Important to this conclusion is the assumption that profits of the GPT sector derive from application sectors that choose to adapt.

In the last chapter Murphy, Romer, and Riddell examine Canadian and U.S. data on wages and the supply of skilled and unskilled workers in order to evaluate the extent to which these data are consistent with the hypothesis that biased technological progress raised the relative productivity of skilled workers with no major discontinuities. They focus on workers with a university degree (skilled) and workers with a high school diploma (unskilled). The stylized fact is that the relative wages of skilled workers increased sharply in the United States for more than two decades, starting in the late 1970s, while the relative supply of skilled workers also increased. In Canada, on the other hand, during the same period the relative wages of skilled workers did not increase while their relative supply did.

From a theoretical point of view these trends can be explained by skill-biased technological progress. Skill-biased technological progress raises the relative demand for skilled workers and thereby raises their relative reward for given employment levels of the two types of labor. On the other hand, a rising relative supply of skilled workers reduces their relative reward. Therefore in a country that faces skill-biased technological change and a rising relative supply of skilled workers the wage premium may increase or decrease, depending on whether the force of technical change dominates the relative supply trend. Since in Canada the relative supply of university graduates increased much faster than in the United States, it follows that the differing Canadian and U.S. experiences do not contradict this theory.

The question is, however, whether the quantitative nature of the data is also consistent with the theory. Murphy, Riddell, and Romer ask this question in the context of two additional restrictions. First is whether the same model fits both the U.S. and Canadian data, and second, whether these data can be explained with no break in the rate of skilled-biased technological change. They conclude that the answer to both questions is affirmative.

These conclusions may be interpreted as bad news for the GPT hypothesis. Two points need to be observed, however. First, this evidence suggests at most that no jump occurred in the relative productivity of skilled workers, but it does not exclude the possibility of a discontinuous technological improvement that was close to neutral with respect to the two types of labor. Second, although discontinuities are possible outcomes of the arrival of a new GPT, they are not inevitable.

The study of GPTs is relatively new. The chapters of this book provide much useful insight to anyone seriously interested in the subject. And they may prove to be foundations for further fruitful work on this important topic.

References

Aghion, P., and P. Howitt. 1992. A model of growth through creative destruction. *Econometrica* 60: 323–51.

Bresnahan, T., and M. Trajtenberg. 1995. General purpose technologies: "Engines of Growth." *Journal of Econometrics* 65: 83–108.

David, P. 1991. Computer and dynamo: The modern productivity paradox in a not-too-distant mirror. In *Technology and Productivity: The Challenge for Economic Policy*. Paris: OECD.

Du Boff, R. B. 1967. The introduction of electric power in American manufacturing. *Economic History Review* 20: 509–18.

Grossman, G. M., and E. Helpman. 1991. *Innovation and Growth in the Global Economy*. Cambridge: MIT Press.

Kuznets, S. 1966. *Modern Economic Growth*. New Haven: Yale University Press.

Landes, D. 1969. *The Unbound Prometheus*. Cambridge: Cambridge University Press.

Mokyr, J. 1990. *The Lever of Riches*. New York: Oxford University Press.

Romer, P. M. 1990. Endogenous technological change. *Journal of Political Economy* 98: S71–S102.

Rosenberg, N. 1982. *Inside the Black Box*. Cambridge: Cambridge University Press.

Rosenberg, N. 1996. Uncertainty and technological change. In R. Landau, T. Taylor and G. Wright, eds., *The Mosaic of Economic Growth*. Stanford: Stanford University Press.

von Tunzelmann, G. N. 1978. *Steam Power and British Industrialization to 1860*. Oxford: Clarendon Press.

2 What Requires Explanation?

Richard G. Lipsey, Cliff Bekar, and Kenneth Carlaw

To get at a working definition of a general purpose technology, we first summarize the state of the theoretical literature. We next look at illustrative historical examples, and using current theories in conjunction with the historical record, we identify a set of technological characteristics that define a GPT. Finally, we check our concept against additional examples and raise the question of its usefulness.

In the course of our discussion, we enumerate what seems to us to be the most relevant of the rich set of facts and empirical generalizations that have been established by students of technological change. We hope these will play the same role that Kaldor intended for his stylized facts in macro growth theory, forcing theories out of the infinite number of spaces derivable from pure conjecture and into a finite number of empirically relevant spaces. It would be utopian to expect a theory to explain all of the awkward facts, but it is desirable that the theories we take seriously do not blatantly conflict with any of them. (This is by no means a weak requirement.)

In the first two sections we consider theories that attempt to capture stylized versions of GPTs in rigorous formulations. We then consider theories that are appreciative in Nelson's (1995) sense of the term. These theories are able to include more of the observed empirical richness of GPTs than can formal theories, but at the cost of being unable to model them matematically.

For comments and suggestions we are indebted to the members of the Canadian Institute for Advanced Research's Economic Growth and Policy Group to whom an earlier version of our chapters were presented and to Peter Howitt, Elhanan Helpman, Nathan Rosenberg, and an anonymous referee for comments on a later version. Although we have used many sources, our largest single debt for facts and insights is to Nathan Rosenberg.

2.1 Formal Theories

Bresnahan and Trajtenberg

In their seminal article on the theory of GPTs, Bresnahan and Trajtenberg (BT) (1992) argue that technologies have a treelike structure, with a few prime movers located at the top and all other technologies radiating out from them. They define GPTs as having three key characteristics: pervasiveness, technological dynamism, and innovational complementarities. Pervasiveness means that a GPT is used in many downstream sectors because it provides a generic function, such as rotary motion. Technological dynamism results from its potential to support continuous innovational efforts and learning, which allows for large increases in the efficiency in the GPT over time. Innovational complementarities exist because "... productivity of R&D in the downstream sectors increases as a consequence of innovation in the GPT, and vice versa."

The consequences of improving the GPT are reduced costs in the downstream application sector, the development of improved downstream products, and the adoption of the GPT in a growing range of downstream uses. The decision to improve the GPT induces more innovational effort in the final applications sectors, which in turn induces more innovation in the GPT. This vertical complementarity causes a nonconvexity in the implicit production function for R&D, which creates a coordination problem between the GPT sector and the application sectors. The nonconvexity is viewed as reflecting a number of real world phenomenon that surround innovation, including information asymmetries, sequencing of innovations, technological uncertainty and coordination problems.

The single immortal GPT is owned by a monopolist who optimally chooses how fast to improve its general productivity. The GPT's users choose how fast to improve their specific application. The authors define a continuum of partial equilibria running from the myopic, noncooperative equilibrium, with a low rate of innovation, to a perfectly coordinated equilibrium. In the latter, all complementary relations are recognized and the rate of innovation is socially optimal.

Helpman and Trajtenberg

In the papers reprinted in chapters 3 and 4 of this volume, Helpman and Trajtenberg (HT) extend BT's paper by modeling a version of the "technology tree" using an explicit general equilibrium framework, by tracking the effects of a new GPT on macro aggregates, and by explicitly modeling the new GPT's diffusion to capture the horizontal externalities that BT discuss.

At any one time there is only one GPT in use, and it is employed only by the sector producing final goods. An R&D sector invents supporting components that are used alongside the GPT, and a third sector produces them. The GPT's productivity depends on the number of these supporting components, which are specific to one GPT. They are modeled in a refinement of the production function found in Grossman and Helpman (1991), who based it on the Dixit-Stiglitz utility function which was developed to model monopolistic competition. This function has the property that as new supporting components are developed, total output increases while productivity per component falls, imposing a finite limit to the GPTs development. It also implies a vertical complementarity between the GPT and its supporting components, and a horizontal substitutability among the supporting components themselves.

Each GPT arrives exogenously. It is immediately recognized as a GPT, and resources are diverted from final production to R&D, which develops the new GPT's supporting component. (If R&D is still going on to develop components for the old GPT, this activity stops immediately, and these resources also move to developing components for the new GPT.) The diversion of labor out of component production, where it produces monopolistically competitive rents, into R&D, where it produces no rents, causes what HT call an "output slowdown." In their model this is a fall in measured output. (It is not clear what happens to total factor productivity, since the assumptions needed for its precise calculation are violated by the existence of a monopolistically competitive sector.) A second possible cause of their output slowdown is a mismeasurement of the full value of the new R&D that makes no immediate contribution to final goods production. Eventually enough components are developed for the productivity of the new GPT to exceed that of the old, and final goods producers switch from the old to the new GPT.

In their second paper HT model the process of a new GPT diffusion. There are many sectors that may potentially adopt the GPT, and each sector has a different productivity in using it. Each sector develops components for the GPT in sequence by diverting resources from production to R&D, starting with the one that has most to gain. (The free-rider problem is ruled out by assumption in both models.) Once it has made its initial component, each sector waits until the next to last phase of the economy's R&D is completed; then all sectors rejoin the R&D process to complete the final phase. Thus the transitional pattern from one GPT to another is determined by the sequencing of the R&D, not the diffusion of the GPT.

Aghion and Howitt
Aghion and Howitt (this volume, chapter 5) provide their response to two empirically relevant issues in HT (1994), which is reproduced here as chapter 3. The first issue concerns the timing of slowdowns that occur immediately after the emergence of the new GPT. Aghion and Howitt argue that this is inconsistent with David's (1991) observations that it may take several decades for a major new technology to have a significant impact on macroeconomic activity. They interpret this to mean that there should be an initial period in which the macro data are unaffected by the arrival of a new GPT.

Aghion and Howitt argue that measurement error and complementarities are two of the three reasons why we should expect to see this period of "no action". The third is social learning. To model this third reason, they adapt most of HT's framework. The new feature is that each sector must develop a specific intermediate good before anyone in that sector can profitably make components for the GPT. After the GPT arrives "serendipitously," firms must first develop sector-specific "templates" using fixed endowments of specialized labor that has no other use. During this phase no resources are moved and nothing changes in measured aggregates. A model of epidemic diffusion generates the desired social learning dynamics. Initially every sector's specialized labor engages in R&D to acquire its template. Success occurs with an initial low probability that increases as more and more sectors acquire templates. When each sector has its own template, resources are moved out of production into R&D to create components for the GPT. When a sufficient number of components have been created, the economy switches over to using the new GPT in production.

Aghion and Howitt also address HT's explanation of the fall in measured output, arguing that the maximum size of the slowdown, based solely on a reallocation of labor to the R&D sector, could not be enough to account for observed slowdowns. Instead, the adjustment and coordination problems associated with the introduction of a new GPT, and its accompanying higher rates of innovation, might increase the rate of job turn over and accelerate the rate of obsolescence. They deal with the unemployment that would result from increased job turnover using a revised version of their main model described above. They deal formally with the obsolescence problem using a revised version of the model presented in Howitt's measurement paper (this volume, chapter 9). In this model there is a range of parametric values for which the increase in the rate of obsolescence, caused by a higher rate of technological change, can lead to a slowdown in measured output.

Dudley
Dudley (forthcoming) follows Innis (1951, 1972) and Dudley (1995) in treating information and communication technologies (ICTs) as the fundamental technology from which all others flow (although he does not use the term GPT). His ICTs have three basic characteristics: to store, to transmit, and to reproduce information. Each ICT is better at some of these functions than others, and R&D increases that strength relative to the others as time goes by. Eventually the existing ICT fulfills its least efficient function too inefficiently to be tolerated. Research switches to the development of a new ICT as an endogenous response. Until the new ICT is well enough developed to be widely adapted, measured productivity growth will be slow (since all R&D is devoted to the sector producing the new ICT, which has a small weight in overall output). This whole process gives rise to an endogenous cycle of ICTs, each better at one of its three functions than the other two. Each cycle is associated with a transitional slowdown. Dudley uses this shift from one type of ICT to another to explain broad historical trends and cycles over 1,000 years.

2.2 Appreciative Theories

Although none of the authors considered in this section use the term GPT, they employ concepts that are similar enough to be of interest to those seeking to understand the phenomenon of pervasive technologies.

Freeman, Perez, and Soete
Several authors have used the concept of a technoeconomic paradigm (TEP), which is much broader than a GPT, since it covers the entire economic system that surrounds any set of pervasive technologies actually in use (e.g., see Freeman, Clark, and Soete 1982; Freeman and Perez 1988; Freeman and Soete 1987). A TEP is a systemic relationship among products, processes, the organizations, and the institutions that coordinate economic activity. A typical paradigm is based on a few key technologies and commodities that are mutually reinforcing, a few key materials whose costs are falling over time, a typical way of organizing economic activity, a typical supporting structure, a typical pattern of industrial concentration, and a typical pattern of geographical location. Although all the elements of a TEP are assumed to be systematically related, the driving force for change comes from what we would call a new GPT, some innovation that fundamentally alters the relationships among various technologies and between technology and the other elements of the TEP.

The concept of a TEP has been used for two major purposes. The first is to argue that changes in important technologies induce structural changes across the whole economy. The second is to develop a theory of the long cycle in which the prevailing technology eventually runs out of scope for improvement, causing severely diminishing returns to further R&D and investment. This crisis in the old TEP provides the endogenous incentive for the development of new technologies that form the core of a new TEP.

Mokyr
Mokyr (1990) considers two types of invention. "Micro inventions" are incremental in nature, largely improving existing technologies and responding to economic incentives, and "macro inventions" are defined as "inventions in which a radical new idea, without clear precedent, emerges more or less ab nihilo." They "... do not seem to obey obvious laws, do not necessarily to respond to incentives, and defy most attempts to relate them to exogenous economic variables" (p. 13).

Mokyr's macro inventions serve two purposes that are relevant to a discussion of GPTs. First, without macro inventions the growth process would eventually come to a halt. Second, because they possess widespread complementarities, they provide a fertile ground for many supporting micro inventions.

Lipsey and Bekar
Lipsey and Bekar (1995) study what they call "enabling technologies," which are defined mainly by their extensive range of use and their complementarities. Enabling technologies are roughly similar to GPTs, although the authors point to fewer classes and fewer cases within each class than we now accept as GPTs. Using historical and current evidence, they argue two main points. First, the introduction of such technologies have in the past caused major changes in the entire structure of the economy, as well as in its economic performance, changes that they call deep structural adjustments, or DSAs. Second, over the last two decades the industrialized economies have been going through such a period, this time caused by the current revolutions in two enabling technologies, made-to-order materials and computer-based ICTs. They also argue that the introduction of a new enabling technology is neither necessary nor sufficient for DSAs. It is not necessary because some less pervasive technologies have caused such economywide repercussions. It is not sufficient because some enabling technologies have not been associated with such wide-ranging repercussions.

2.3 GPTs in History

If the concept of a GPT is to be useful, then GPTs must be identifiable. Our procedure is first to choose a set of important technologies according to their observed economic effects. We search through history to find examples in which a widely used new technology caused changes that pervaded the entire economy, Lipsey and Bekar's DSAs. As a second step we look for common technological characteristics which we use to define a GPT.[1]

The introduction of a major new technology whose effects reverberate through the entire economy, sometimes affecting the political, economic and social structures, is a relatively rare event. To get clear and dramatic examples of technologies which we want to include in our set of GPTs, we have ranged as far back in economic history as the neolithic agricultural revolution. Although the documentation of our earlier examples is less detailed than those from more modern times, we have selected only cases where the broad outlines of what happened are clearly established by historical and/or archaeological research. While the effects of many of the GPTs we examine may seem extreme to some readers, every one of those that we mention is supported by strong empirical evidence that is not easily appreciated until one looks at the historical record in some detail. Of course many of the GPTs only contributed to the outcomes that we mention, as opposed to being their sole cause. After choosing our examples, we found that they all fell within four main activities that are covered in the rest of section 2.3. Although we have not tried to be exhaustive, we believe we have identified most of the historical cases where new, pervasive technologies caused profound DSAs. Below we briefly allude to most of these, although space has caused us to delete our discussion of iron, steel, and several major transport technologies.

2.3.1 Information and Communication Technologies (ICT)

Providing, analyzing, and using information is essential to all coordinated economic activities.

---

1. We believe that this is what one does, but less explicitly, whenever one develops a theory about some new phenomenon. The phenomenon is observed, and we study it to see what seem to be its key characteristics. We then formalize a stylized vesion of these in a set of assumptions designed to capture the phenomenon (a process that Blaug 1980 calls adduction). If we define useful concepts we are able to develop a viable theory.

Writing

Before the invention of symbols to represent the spoken word, records were mostly held in human memory and communicated orally. (One important exception was the symbols for recording quantities that were the precursors of writing.) Writing was invented independently in several places, allowing information to be communicated accurately over time and distance, and leading to a major increase in the complexity of economic activities. Its invention in Sumer around 3,500 BC was accompanied by the development of sophisticated systems of taxation and public spending—systems that were quite impossible when most records were held in memory. The new public savings were largely used to finance the building of irrigation works, whose technology evolved rapidly. In Sume the area under cultivation increased, and agricultural surpluses rose. The populations of the largest settlements, which had been measured in the hundreds for millennia, increased over two centuries into the tens of thousands. A division of labor appeared among the priesthood whose members raised the taxes and supervised the first irrigation systems. Temples, larger and more complex than anything seen before, made their appearance.

In short, there was a radical transformation of the societies of the Tigris–Euphrates valley. The changes in the two or three centuries following the widespread use of writing probably exceeded the changes over the two or three millennia preceding it. While it is unlikely that all of these changes could be attributed directly to writing, the evidence does suggest that most of them would have been difficult, and some of them impossible, to accomplish in a society without written records (Dudley 1991, ch. 1; Schmandt and Besserat 1992).

Printing

Economic, political, and social historians agree that the widespread use of printing contributed in important ways to the dramatic changes that occurred in Europe between 1500 and 1700. With manuscripts, the major cost of duplications was the variable cost of the scribe's time; with printing, the major cost was the fixed cost of typesetting, and the marginal cost of printing an extra copy was small. This new cost structure made mass communication feasible.

At first, printing achieved a low-level equilibrium typical of new technologies with network externalities (Arthur 1988). Texts were printed in Latin for the few who could read. To escape this equilibrium, two changes were needed. First, a relatively large amount of printed material had to be published in the popular vernaculars. The printers and grammarians of Europe played a key role in developing and implementing standardized version of each vernacular over the century following the introduction of printing. Second, a significant proportion of the population had to be trained to read the vernaculars. Here the Protestant revolution was critical. Protestant thinkers insisted that their followers be able to read and interpret the scriptures for themselves. Mass communication helped the Protestant revolution, since its direct appeal to the people would have been impossible without the many low-cost printed pamphlets written in the vernacular.

As these developments occurred and the costs of large-scale publication fell, learning exploded. Monopolies of knowledge were upset. The Netherlands were tolerant of both printing and the new knowledge that it represented. Thinkers, writers, and printers, exiled from other parts of Europe, moved there. The creation of the Dutch information network based on low-cost reproduction of the printed word greatly increased productive efficiency and tax revenues. Although many factors contributed to their eventual victory over Spain, and their rivalry with Britain as a world trading power, research indicates that a key part of their success was owed to their liberal attitude toward the technology of printing and the learning that it embodied (e.g., see Dudley 1991; Cipolla 1993; Cardwell 1972; Huff 1993).

Current ICT Revolution
The latest ICT revolution is being driven by the electronic computer in its many and varied forms. Our account of it is brief, since most of the dramatic changes are well known and are referred to elsewhere in this volume. (For one excellent succinct survey, see Greenwood 1997.) The revolution is changing product design, production, marketing, finance, and the organization of firms. It is also creating a wide range of new products incorporating hard coded chips, computers, and/or software. Computers are used to fly aeroplanes, drive trains, operate machines, run buildings systems, warn of unsafe driving practices, monitor health, to facilitate communication through the interned, e-mail, and desktop publishing. By managing information flows more effectively than did the old, hierarchally organized, mass of middle managers, computers are causing major reorganizations in the management of firms. Labor productivity is increasing rapidly in many of the affected industries, even if it has yet to show up strongly in the aggregate data.

2.3.2 Materials

Materials are required for all consumers' goods and for all process technologies. Few services are provided by unaided labor, and most complex service operations require elaborate process technologies that are embodied in physical capital. The very concepts of the stone age, the bronze age, the iron age, and the age of steel highlight the importance of materials and the structural transformations caused when one pervasive materials technology succeeds another.

Bronze

Stone, clay (pottery), and wood were the universal materials before the age of metals. Bronze gave its name to an age lasting for about fifteen centuries, starting about 2,800 BC. The invention of bronze, the first material malleable enough to be worked but strong enough for most uses, facilitated a vast range of new technologies for both civilian and military uses. These led to many wide-ranging social and political changes.

Prior to bronze, the main external threat to town dwellers was attack by uncoordinated bands of migrant barbarians. With bronze weapons, trained armies and multiple city empires appeared for the first time, since bronze weapons and interlocking bronze shields gave enormous scale advantages to large armies over smaller ones. Cities now became walled against the main external threat of attack by well-equipped and disciplined armies.

Once the optimal size of the state grew to cover a wide geographic area, it became too expensive to coordinate transactions through centralized authority. As a result the command system (documented to have been a major allocator of resources in compact single-city theocracies) was largely replaced by markets. Growing markets facilitated new divisions of labor. A bronze age empire of 2,400 BC was demonstrably more complex in economic, technological, and social dimensions than a stone age town of 3,000 BC. The reason, at least in part, is the introduction of the workable metal for which the age is named (Dudley 1991; Drews 1992).

Made-to-Order Materials

New materials became increasingly important after the growth of the chemical industry at the end of the last century. Initially a new material was invented in isolation and used to substitute for some existing material. Today, however, materials are invented specifically for the new products and processes that are being developed. New materials are seen as crucial to the continued expansion of many important growth sectors, including microelectronics, transportation, architecture, construction, energy systems, aerospace engineering, the automobile industry, and, to look further into the future, fusion reactors, ersatz human organs, and solar conversion cells. The technology of made-to-order materials is creating clusters of related innovations in often widely differentiated industries:

Changes in materials innovation and application within the last half century ... have occurred in a time span which was revolutionary rather than evolutionary. The materials revolution of our times is qualitative as well as quantitative. It breeds the attitude of purposeful creativity rather than modification of natural materials, and also a new approach—an innovative organization of science and technology. (Kranzberg and Smith 1988, p. 88)

2.3.3 *Power Delivery Systems*

Virtually all productive activities require the application of power. It is useful in some cases to distinguish between the primary generator of the power and the system that delivers that power. For example, electricity was the first power system to separate the place of generation from the place of use by more than the distance allowed by such physical links as belts and shafts. This separation had important consequences. Our analysis is not, however, affected by the distinction, so we treat the power source and its delivery mechanisms as a single technology, which we call "power delivery systems."

Waterwheel
During the medieval period the spreading use of the waterwheel led to the mechanization of European manufacturing. The power of animals and humans (often slaves) was largely displaced by the wheel, setting Europe on a trajectory of mechanization which eventually took its industry well beyond that of Islam and China.

The waterwheel went through many improvements. Furthermore important ancillary inventions, particularly the cam, which turns rotary motion into linear motion, expanded its range of activities to include sawing, hammering, grinding, pumping, beer brewing, tanning, papermaking, and cloth fulling.

During the Middle Ages the use of water power spread to include the development of floating mills, large dams, and mills that harnessed tidal power. These were capital intensive and were financed by some of the earliest share issues. Also new property rights had to be developed to regulate the use of the river for power (Gimpel 1967; Gies and Gies

1994). Technological improvements continued through the nineteenth century when water wheels were challenged by steam and finally eliminated by electricity in the late nineteenth and early twentieth centuries.

Steam

Commercial steam power started with Savery's atmospheric engine. Although technically inefficient and dangerous, it was quite widely used because it outperformed the available alternatives, all of which used animate power. In 1712 Newcomen invented a crude but effective and relatively safe atmospheric steam pump which was soon widely used in coal mines. Micro improvements alone almost doubled the efficiency of the Newcomen engine. James Watt transformed Newcomen's engine into one in which steam did the driving rather than atmospheric pressure. Although he greatly improved the efficiency of his new reciprocating engine over the years, he did not believe in high pressure. Only after his patent expired in 1800, did the development of the high pressure engine allow steam to expand into many new uses especially in transportation. High-pressure engines required strong materials, which caused many improvements in metallurgy. Taken as a whole, late nineteenth-century steam engines enjoyed many orders of magnitude increases in efficiency over their early eighteenth-century counterparts (Landes 1969; Von Tunzelman 1995).

Steam gradually replaced water power as the main source of industrial energy throughout the first half of the nineteenth century. At first, steam worked within factories designed for water power. Later, however, new factories were designed to suit the steam engine. The economies of scale associated with steam power made large factories more efficient than smaller ones. Locational advantages altered drastically, since factories were no longer tied to fast-moving water. The total amount of power available to industry increased many fold over water-powered factories. As is typical of the competition between old and new technologies, systems based on water power fought back, and a series of technological advances improved their efficiency through the early part of the nineteenth century. Right up to the end of the century, water was used to power many factories, particularly in the textile industry.

Steam allowed for the creation of new products, new production techniques, and, eventually, many new industries. Steam engines slowly replaced sails at sea and rapidly replaced long-distance horse transport on land. The railways transformed the patterns of industrial location that had resulted from road and water transport, opening up new areas for the growing of grain and significantly lowering the price of food.

Electricity

The dynamo converts mechanical power into flows of electrons that have multiple uses. It is probably the most pervasive energy delivery innovation of all time.

Electricity started with only a small number of applications. Slowly, as technical problems were solved, the number of uses expanded, transforming the techniques and locations of production and leading to a range of new products and industries.

Electricity powers factories and lights cities. It has spawned a range of new consumer's durables. For example, an assortment of electrically driven household machines including washing machines, dishwashers, vacuum cleaners, irons, refrigerators, deep freezers, and electric stoves transformed household work and eliminated the battery of servants that was required to run middle-class households in 1900. Electricity has powered an ongoing communications revolution, starting with the telegraph which, for the first time in history, provided a publicly available system that allowed information to travel faster than human messengers. The new communications technologies made possible by electricity evolved through the telephone, the radio, and the TV. Electricity also powers the computer which is the basis of the current ICT revolution. Electricity is therefore complementary with the new computer-based GPT.

The full development of electricity's potential required substantial structural alterations across the entire economy. One of the most important was a drastic alteration in the layout of factories. Water and steam, used a central drive shaft whose power was distributed throughout the factory via a set of pulleys and belts. Because of heavy friction loss in belt transmission, machines that used the most power were placed closest to the drive shaft, and factories were built with two stories to get more machines close to the shaft. At first, electric motors merely replaced steam or water as the power source for the central drive; they were installed in a design adapted to the old power sources. Later, a separate motor was attached to each machine (the unit drive), after an intermediate stage of group drives. It was then slowly realized that the factory could be built as a single story and the machines arranged in the order of the flow of production. Only when the restructuring was completed, was the full potential of electric power in factories realized (Schurr 1990; David 1991).

Electricity altered scale economies. In some lines of production, particularly assembly, scale economies were increased; in others, however, small-scale production became more cost efficient because an electric motor could be attached to each machine tool. The result was a system of small decentralized parts producers supplying large centralized assembly plants—a method of production that is still used today. The 1890s were also a time of intense merger activity, which was sometimes the cause, and sometimes the effect, of electrification (Chandler 1990).

Internal Combustion Engine

The development of the internal combustion engine overlapped the later stages of steam and the early stages of electricity. It began as a relatively inefficient atmospheric engine in which the explosion of coal gas pushed a cylinder upward and atmospheric pressure pushed it down on its power stroke. Soon, however, the explosion of gas was used for the power stroke, and by setting the cylinder on its side, a quiet, efficient, highly successful four-cycle engine was produced. A decade passed before the technical difficulties of using gasoline were overcome and the engine became mobile. Multiple uses quickly developed. The automobile was one of the first to use it. The engine's relatively low weight/horse-power ratio also overcame one of the most important obstacles to powered flight by heavier-than-air craft. The gasoline powered internal combustion engine thus enabled two of the twentieth century's most important transportation GPTs.

As the design and efficiency of the internal combustion engine improved, its range of use increased. Its weight/power ratio fell far enough to allow its use in lawn mowers, power saws, small commercial aircraft, forklifts, and anywhere else that a lightweight, dependable, mobile power source was required (Cardwell 1995, pp. 338–49).

2.3.4 *Transportation*

All goods production requires that something, be it raw materials, semi-finished, or final goods, be transferred over space. We confine ourselves to the railway and the motor vehicle, omitting several other transport technologies that were revolutionary in their time and had the type of effects in which we are interested, such as aircraft, the iron steam ship, the three-masted sailing ship, rowed galleys, the original use of sails, and the wheel.

Railways

The railway was enabled by the steam engine, the first mobile, inanimate power delivery system for land uses. (Waterwheels were tied to rivers, while wind was unsuitable as a mobile source of power on land.) Over time the railway largely replaced canals, as well as the horse and wagon for long- and medium-distance travel. In its turn it was challenged by the motor vehicle, but this competition did not lead to its elimination, only to its restriction to a smaller market. Even today railways remain an efficient way of transporting goods with low value/weight ratios over long distances and of transporting people over the kind of middle distances that account for much European travel.

The railway, in conjunction with the iron steamship and the telegraph, opened up vast parts of the world to settlement and allowed the production of foodstuffs for distant markets. This group of technologies was also the foundation of a growing tourist industry. Telegraphic communications permitted faster speed and required more complex forms of organization, which later spread from railway companies to other industries (Billington 1996).

Motor Vehicle

The motor vehicle is a multiple purpose technology that has profoundly affected the twentieth-century economy. The truck in all its forms is a commercial and military transport vehical (armored it became the armored military car and tank, with momentous repercussions). The automobile is used for transport to work and shopping, for recreastion, sport, and countless other purposes.

Among the many structural adjustments caused by the automobile were important changes in the way people lived. It reinforced the move to the suburbs initiated by suburban electrical railways. Along with the refrigerator it helped replace the dominant small grocery store by the supermarket, which gave rise to the suburban shopping center. Commercial trucking challenged railways over long and intermediate hauls and eliminated the horse and wagon for short hauls. Next the U.S. interstate superhighway system contributed to the movement of manufacturing from locations in the inner city near the rail head to the suburbs alongside the new highways, furthering the decay of the U.S. inner cities.

The automobile industry was profoundly affected by the development, early in the twentieth century, of machine tools that could cut prehardened steel. Because the new machine tools allowed a high degree of accuracy in production, parts with identical engineering specifications became identical in practice. Interchangeable parts made many of Henry Ford's early process innovations possible. It was a short step to add the conveyor belt that typified mass production techniques in the public's mind, but that was only the last step in a series of technological innovations following from prehardened parts (Womack et al. 1990). These developments spelled the end of the artisan in many lines of North American production and produced the highly paid, but relatively unskilled, worker doing repetitive work on dedicated machinery.[2] Numerous union practices, such as job demarcation, came in its wake. Many of the important changes taking place in manufacturing today, such as lean production and just-in-time systems, had their start in the automobile industry, which has been a constant innovator of new production techniques.

2.4 Technological Interrelatedness

Interrelations among technologies play an important part in determining the overall reaction to specific technological changes. We call the technology that specifies each physically distinct, stand-alone, capital good a "main technology."

First, we note that most main technologies have differentiated parts. For example, a commercial airliner is made up of a large number of subtechnologies including an engine to deliver power, a thrust technology to turn that power into movement, a body, an undercarriage, a navigation system, and an internal control system. Analysis of these subtechnologies shows them to be made up of other subtechnologies. An aircraft's navigation system is composed, for example, of compasses, gyroscopes, computers, sensing devices, radios, radar, and so on. Analysis of each shows them all in turn to be made up of other smaller subtechnologies. Notice that some of the subtechnologies, such as compasses and food warmers, can also be used on their own as main technologies. In contrast, other subtechnologies, such as the aircraft's rudder or windows, are useful only as a part of this specific main technology. This fractal-like nature of the aircraft's makeup is typical of virtually all capital goods. It is also typical of consumers' durables, such as automobiles and refrigerators, that deliver services for use in consumption. The interdependence of the subtechnologies is often Leontieff in nature; the main technology will not function if you remove one of its subtechnologies. For example, a standard gasoline engine will not run without its spark plug. Other subtechnologies increase

---

2. Ford did not invent the assembly line used, among other places, in the manufacture of firearms in the nineteenth century and in the production of Venetian galleys prior to the battle of Lepanto in 1574. But by insisting on standardized parts for his automobiles, he extended its use to mainline manufacturing.

the efficiency of the main technology without being essential. For example, the air filter is not necessary for the internal combustion engine, but the engine's efficiency is much increased by it. Each generic type of subtechnology, such as a spark plug or a tire, usually comes in several differentiated versions that are close substitutes for each other.

Second, we observe that main technologies are typically grouped into *technology systems*, which we define as a set of two or more main technologies that cooperate to produce some range of related goods or services. They cooperate within one firm, among firms within one industry, among firms in closely linked sets of industries, and even across industries that are seemingly unrelated from an engineering point of view (Rosenberg 1976, 1982). Technology systems overlap each other in the sense that a subset of the technologies that are used to produce product A is used, along with other technologies, to make product B, and so on. In some cases the technology may be used in the process technology that produces the good that embodies it, as when computers are used to manufacture computers. In other cases some main technologies assist the operations of others without being necessary. For example, much of the value of a computer depends on its cooperation with its peripherals, such as printers, modems and software. Notice that many of the separate main technologies in some technology systems may simultaneously compete with, as well as complement, each other. Trucks deliver freight to the railhead, making railways more useful and profitable, but they also compete with rail for long-distance haulage.

Notice the effects of a technology system cannot be correctly measured by estimating the consequences of introducing each main technology in the absence of the others. Because of complementarities the effect of the whole is substantially larger than the sum of the effects of the parts. One excellent example is the technology system created by the development in the nineteenth century of the railroad, the iron steamship, the telegraph, and refrigeration. Individually each would have been important. Acting together, they globalized the markets for many agricultural commodities with many resulting DSAs.

Third, technological interrelationships occur in the vertical relations among industries when the output of one is used as an input by another. Materials-producing industries, such as iron and steel, forest products, and aluminium, create inputs used in manufacturing.

Fourth, some industries that produce different and unrelated outputs use similar process technologies—a phenomenon that Rosenberg (1976) calls "technological convergence." This facilitates discontinuous jumps in product technologies because each such product may not have to develop its own radically different process technology. For example, early in the twentieth century the new aircraft industry used process technologies that were already well developed by the bicycle and sewing machine industries. (For a formal treatment of this point, see Bresnahan and Gambaradella, chapter 10 this volume.)

In summary, the economy's overall technology system is a set of interlocking embodied technologies. First, there is the fractal-like web of lower-level, subtechnologies that form any one main technology. Although some of the subcomponents in this engineering structure are necessary, others are important auxiliaries but not essential. Furthermore each general type of subcomponent often comes in various versions which compete with each other. Second, there is an external structure of interrelations across capital goods in one industry, as when several capital goods cooperate to produce a final product. Third, there are interrelationships across industries, as when the output of one industry is used as inputs in another. Fourth, there are process interrelationships when technologically distinct products are produced by technologically similar processes.

2.5 Defining a GPT

Our view of definitions is nominalist not essentialist; definitions are not judged as being right or wrong but only as being helpful or unhelpful in delineating useful categories. We have now identified a group of GPTs by their impacts on the economy. However, if we want to develop theories that predict the effects of GPTs, including some that we have not yet identified and some that do not yet exist, we cannot define them by their effects. We thus search for a definition stated exclusively in terms of technological characteristics. Our definition must be wide enough to include all of those technologies we have identified above, and narrow enough to exclude other less important technologies, such as elastic bands and screw drivers.

2.5.1 *What a GPT Is Not*

We begin by discussing some defining characteristics that have been suggested by others but which we argue are neither necessary nor sufficient to identify the technologies on our list.

Not Defined by Its Demand Characteristics[3]

Schmookler (1966) presents strong evidence that the amount of innovation in an industry, and its characteristic logistic curve of productivity development, are strongly influenced by demand. This is important evidence against those who argue that innovation is solely determined by technological possibilities. GPTs are no exception. A technology will not become a GPT unless it fills numerous demand niches; some will already be filled by existing technologies, some will already be perceived but only inadequately filled, and some will be created by the new GPT itself. Although this broad-based pattern of demand is a necessary condition for a new technology to evolve into a GPT, it is not part of its definition, which depends on its technological characteristics alone. Just as an adequate food intake is a necessary condition for an infant to grow into an adult but not a part of the definition of an adult, so a broad-based demand is a necessary condition for a new technology to evolve into a GPT, but not a characteristic that enters into its definition. This illustrates the general point that the conditions necessary for the development of the characteristics that define some object are not part of that object's definition.

Not Defined in Terms of a Generic Function

Both Bresnahan and Trajtenberg (1995) and Helpman and Trajtenberg (chapter 3) make a defining characteristic of GPTs the provision of what they call a generic function which they illustrate by rotary motion. Webster's dictionary defines generic as "referring to a whole kind, class, or group." Thus the literal meaning of a generic technological function would be a function common to a class of technologies. For example, a generic function of all modes of transport is to move things about spatially, while a generic function of all toothpastes is to clean teeth. In this sense all technologies in some class must provide the generic functions that define the class. Taken in this sense, providing a generic function is not *sufficient* to make a technology a GPT for at least two reasons. First, the generic function itself may be relatively unimportant, such as cleaning teeth. Second, both important and unimportant members of some given class of technologies will provide the generic functions that define their class. For example, rotary motion, a generic function whose provision

---

3. This section was added only after more than one student of technology worried about our neglect of the demand side, for example, quoting Schmookler (1972).

Bresnahan and Trajtenberg use to identify the steam engine as a GPT, is produced by revolving axels or wheels that can be powered by many power delivery systems which we would not wish to include in our category of GPTs such as clockwork motors, the animate power of animals and humans, atmospheric pressure, elastic bands, and tidal forces.

It is also worth noticing that a GPT is not typically characterized by a single generic function that it performs (therefore doing so cannot be necessary). Power delivery technologies, including the waterwheel and steam, provide power for all types of motion. One of the critical inventions in the long history of the waterwheel was the cam which turns the waterwheel's rotary motion into the linear motion to drive hammers, saws, presses, bellows, pumps, and many other important "nonrotary" activities. The principle of the cam acting in reverse also turns the linear motion of the pistons of steam and internal combustion engines into rotary motion where required, while the piston delivers linear motion directly in the many places where that is required. The dynamo does not deliver any sort of mechanical motion. Instead, it delivers electricity which is energy in the useful form of electron flows. Users then transform the electricity into the type of work that they require, including mechanical motion, lighting, heating, impulses for telegraphs, power for switches in the form of vacuum tubes and transistors, power for making aluminum, and so on, each of which may be separately construed as a generic function.

It is possible, however, to reinterpret "genericness" to be a characteristic of a GPT. Consider a class of diverse technologies that use as an input the outputs of some single technology. That single technology many be described as generic to the class that uses it even though it may provide a variety of different functions such as turning a motor, heating an element, or operating a toggle switch. In this sense Bresnahan and Traijtenberg's GPT with a generic function means a technology with many and varied uses, characteristic which we also include in our definition developed below.

Not Always Exogenous nor Always Endogenous

Modern research has left little doubt that technological change is largely endogenous to microeconomic incentives. While technological change is also substantially influenced by pure scientific research, scientific agendas are often themselves strongly influenced by economic signals—as argued by Schmookler (1965) and Rosenberg (1982, 1994). Thus, both research and development are significantly endogenous, responding to the economic signal of profit expectations.

Some historians and theorists have suggested as a matter of fact, or assumed as a matter of theoretical convenience, that new GPTs are exceptions to this rule (e.g., Mokyr 1990; Helpman and Trajtenberg, chapter 3 this volume). Others have argued that GPTs always occur as endogenous responses to economic incentives (e.g., Dudley 1995; Freeman et al. 1982). To proceed with this issue, we need to ask two questions: To what stage in the technology's evolution are we referring? And endogenous to what?

To answer the first question, note that whatever the original source of some new technology, once its potential to create profits is appreciated, its further evolution becomes endogenous. Its R&D rapidly comes to be directed by profit-seeking decisions. So we must be referring to the early stages of a technology's evolution if we are to consider the possibility of it being exogenous to the economic system.

Endogenous to the economic system. In some cases the initial development of a technology that evolves into a GPT is endogenous from the outset. Often it begins as a specialized response to the profit opportunities created by a localized "crisis" in some existing technology. For example, Newcomen's atmospheric engine, and Savery's before that, were endogenous responses to the need to pump water efficiently out of ever-deepening mines, a job that existing technologies could not do effectively.

Exogenous to the economic system but endogenous to science. A new technology is endogenous to pure science when its evolution is being driven primarily by non-profit-related motives associated with a scientific research program. The use of electricity as a power delivery system resulted from discoveries in a seventeenth-and eighteenth-century research program, which was largely driven by scientific curiosity as opposed to the pursuit of commercial gain.[4] Furthermore the technology it replaced, the steam engine, had not reached any obvious limits to its possible improvement.[5]

---

4. Dividing lines are seldom as clear-cut in reality as we would have them in theory. In this case, however, it seems fairly clear that the advances in knowledge about electricity were pretty well exogenous to the economic system right up until the early nineteenth century. When Volta developed his electric battery in 1799, the potential for valuable applications was immediately apparent. Many of the subsequent key discoveries by such scientists as Faraday and Maxwell, however, seem to have been primarily motivated by scientific curiosity. Just when the crossover from mainly exogenously to mainly endogenously driven occurred is debatable. It would seem to be somewhere in the first half of the nineteenth century, possibly around the middle of that period.

5. This seems to us to be an implication of von Tunzelman's detailed 1978 study of the steam engines, although he never says so in so many words.

Thus electricity may best be regarded as exogenous to the economic system, at least until some time into the nineteenth century.

Exogenous to the economic system but endogenous to the political-military system. A technology is endogenous to the political-military system if its evolution is being driven primarily by motives of military or political advantage. The electronic computer was initially developed to do complicated calculations associated with ballistics and code breaking. Although it was developed in universities, the research was funded by the military. It began as a single-purpose machine with very few economic applications. As it was slowly developed to handle different problems, more and more scope for economic uses became obvious and its evolution became increasingly driven by profit-seeking motives.

Technologies of all sorts and sizes have various origins. The GPTs on our list include some technologies that developed as endogenous responses to economic signals and others whose development was exogenous to the economic system. To conform with the awkward facts, theories of GPTs should not make their early evolution either always exogenous or always endogenous to economic signals. A more difficult problem is to explain why GPTs are sometimes exogenous and at other times endogenous.

Not a Radical New Technology without Clear Antecedents
Virtually all major technologies evolve along paths that include both small incremental improvements and occasional jumps. To distinguish these, investigators often define two categories. An innovation is *incremental* if it is an improvement to an existing technology and *radical* if it could not have evolved through incremental improvements in the technology that it challenges in some particular use (Freeman and Perez 1988).

If we are only concerned with the consequences of technological change, this distinction is all that is needed. However, if we are interested in the origins of innovations, some confusions that exist in the literature require us to go further. To do so, we distinguish two separate evolutionary paths and two separate meanings of the term radical. The first path is the evolution of a series of technologies that are applied to *a specific use*, such as the reproduction of the written word. All of the GPTs in our historical list are radical in this sense: They could not have evolved out of the technologies that they challenged and eventually displaced in some particular use. For example, bronze could not have evolved out of stone, nor electricity out of steam, nor the printing press out of quill and ink.

The second path is the evolution of *a specific technology*, such as the printing press. None of the technologies on our historical list, however, are radical in this sense: They did not appear more or less out of the blue. Instead, each had evolutionary paths stretching backward for centuries. For example, the replacement of carved wooden blocks by movable type was an incremental innovation in the technology of printing, which had a long evolutionary history stretching back to its early origins in China. However, from the point of view of technologies used for reproducing the written word, printing was a radical innovation that replaced hand copying. Cases of technologies that are radical in the second sense appear to be much rarer and often are accidental discoveries. Three examples that can be argued as appearing out of the blue without clear parentage are X rays, penicillin, and radio astronomy.

We must now ask if being radical in either sense is a necessary or a sufficient characteristic for a technology to be a GPT. Being radical in the first sense cannot be a *sufficient* condition, since many technologies that we would not want to call GPTs did not evolve out of the technologies they replaced: The nylon stocking did not evolve incrementally from the silk stocking, ball point pens did not evolve out of the fountain pen. We also argue that being radical in the first sense is not *necessary*. For example, the three-masted sailing vessel was a GPT that was not radical with respect to the nautical technologies that preceded it. The key to this innovation was the combination of the square rigged sail, which was used on the cog that the three-masted ship largely replaced along the Atlantic coast, and the lateen sail, which had been used in the Mediterranean for centuries. Yet the two in combination produced a radically new type of vessel that could make transoceanic trips in relative safety—a vessel that was subsequently improved by a series of smaller inventions covering such things as rigging, sails, and provisions. This can be seen as an evolutionary change in sailing technology, a recombination of existing sub-technologies. But its effects were as dramatic as if it had been a completely new technology. It transformed the map of European economic power and led to the first global expansion of European sea borne trade and colonization.

It is easy to see that being radical in the second sense is neither necessary nor sufficient for a GPT, since none of our historical examples are of this out-of-the-blue variety, and some that are out of the blue are clearly not GPTs.

A Confusion about Being Radical Those who debate whether economic growth is continuous or punctuated with occasional discontinuities do not usually distinguish between the two senses of radical. Those arguing for the absence of discontinuities often point out correctly that very few inventions are radical in the second sense of having no clear parentage. The apparent implication is that all innovations must be incremental. The strong suggestion is then that economic growth and technological change will all be continuous and devoid of discontinuities.

Mokyr dissents from this conclusion. He does not, however, distinguish the two senses of radical but argues that what he calls macro inventions are radical in our second sense: They are "without clear precedent [emerging] more or less ab nihilo" (Mokyr 1990, p. 13). He concludes that there can be nonincremental, technological changes and discontinuities in the process of growth.

We agree with Mokyr on the important issue that there can be non-incremental technological changes and big technological shocks to the economy. We argue, however, these are typically initiated by technologies that are radical only in the first sense of not evolving out of the technologies that they challenged. Thus we avoid having to maintain the dubious proposition that our GPTs (or Mokyr's macro inventions) are typically radical in the second sense of being without clear evolutionary histories and appearing more or less out of the blue.

2.5.2 What a GPT Is

All of the technologies in our historical list share four characteristics. They enter as fairly crude technologies with a limited number of uses, but they evolve into much more complex technologies with dramatic increases in the range of their use across the economy and in the number of economic outputs that they produce. As mature technologies they are widely used for a number of different purposes, and they have many complementarities. In the following sections we elaborate on these observations in the following way: We assert that the four characteristics we have just listed are each necessary conditions for a technology to be a GPT. We then use empirical evidence to refute the assertion that any one of them is sufficient. We end by defining this bundle of four characteristics as necessary and sufficient for any technology to qualify as a GPT.

Scope for Improvement
The way in which agents learn about technologies, and the complexity of the economy's whole technology system, implies that any technology that ends up being widely used in many different applications must go

through a process of evolution. Over time the technology is improved, its costs of operation in existing uses falls, its value is improved by the invention of technologies that support it, and its range of use widens while the variety of its uses increases. This is true of both the technologies of the products themselves, such as steel or the printing press, and of the processes by which they are made. Every one of the major technologies discussed in our historical section displays this evolutionary experience in which the processes of technological change and diffusion are intermingled in time, space, and function. Scope for improvement when a technology is first introduced is a necessary characteristic for a technology to become a GPT. Since this characteristic is shared by many technologies that we would not want to include as GPTs (e.g., ploughs, bicycles, and drill presses), it cannot be a sufficient condition.

Wide Variety of Uses
At a high enough level of abstraction, every technology, or class of technologies, produces only one output. For example, we can think of transport technologies as producing only transport, or power delivery systems as producing only energy. However, at a lower level of abstraction, technologies often produce more then one distinct output. For example, the dynamo provides mechanical power for a wide variety of productive activities including electric motors, lighting, heating, telegraphs, transistors, radios, televisions, computers, steel (mini-mills), and aluminium.

Each of the technologies on our historical list are used in a wide variety of products and processes. Furthermore they are used as physical components in the makeups of main technologies and/or as main technologies within technology systems.

Variety of uses is time dependant. A new GPT typically has a few very specific uses, but as it evolves, many applications are discovered. Thus a new technology of this type has implicit in it a major research program for improvements, adaptations, and modifications.

So a necessary condition for a GPT is that at some stage of its development it comes to have a wide variety of uses. This cannot, however, be a sufficient condition, since there are a number of non-GPTs that also share this characteristic. For example, belts act as power delivery systems to, from, and within many machines; they also act as a transport system in assembly plants, in airports, and in loading and unloading transport vehicles. X rays are used in medical imagery, cancer treatment, security, archeology, saw milling, and mineralogy. However, both X rays and belts lack sufficient complementarities and range of use to be included within our definition of GPTs.

Wide Range of Use

By a technology's range of use we refer to the proportion of the productive activities in the economy using that technology. This range runs the gamut from one industry to the entire economy. For example, a post-hole digger is used only in the manufacture of fences, a product not widely used in the economy. Electric lightbulbs and screw drivers are used more or less throughout the entire economy. Notice that having a wide range of use is not the same thing as having a wide variety of uses. For example, although it is widely used across the economy in many different settings, an electric lightbulb has only one use, to produce light. In contrast, lasers feed information to cash registers at store checkouts, provide precision cuts in surgical operations, read information from CDS, and print hard copies from computer outputs.

Many of the GPTs that we have studied, such as electricity and computers, are, or have been, used across virtually all of the economy. Others, such as steam and automobiles, spread across most of the economy. As with number of uses, this is a characteristic that evolves over time. GPTs typically emerge as sector-specific technologies whose use spreads only slowly over the economy. For example, although steam power was initially only important in mining, by the end of its evolutionary trajectory its influence had spread across many sectors.

Notice that having a wide range of use is relative to the part of the economy being studied. We define range of use with respect to the whole economy. Others who study, a sector, or an industry properly define range of use more narrowly. (For example, in chapter 7 of this volume Rosenberg looks at technologies that are GPTs in the chemical industry but might not be GPTs economywide.)

So wide range of use is another necessary characteristic of GPTs, but it cannot be sufficient, since GPTs share this characteristic of widespread use with many other technologies, such as electric lightbulbs and screw drivers which produce a single economic output that is widely used across the economy.

Strong Technological Complementarities with Existing or Potential New Technologies

In standard microeconomic theory, complementarities refer to the response of quantities to a change in price. In contrast, when students of technological change speak of complementarities, they are often referring to the impact of a new technology. Furthermore technological inter-relations are often referred to in the literature as complementarities, although the intended meaning is often "closely related" rather than complementarity in the theoretical sense. Game theory introduces the concept of "strategic" complementarities, where the actions of one agent affect the payoffs of another. In technological competition the most obvious example is when R&D done by A increases the expected value of the R&D done by B. This strategic complementarity covers some, but not all, of what we call complementarities. We reserve the term complementarity to deal with the response of the system to certain types of technological changes and distinguish two types of complementarities, which we call Hicksian and technological.

Hicksian Complementarity The Hicksian concepts of complementarity and substitutability in production theory refer to the signs of the quantity responses to a change in some price. *Net complementarity* is defined for a constant level of output, while *gross complementarity* allows output to change in response to a change in one input price, and thus combines the output effect with the substitution effect.

An innovation that reduces the cost of an input, x, that is widely used as a service flow in many production processes will cause substitutions among inputs. It will also increase the demand for other inputs that prove to be gross complements to x. Where the demands for inputs other than x increase *in response either to an actual fall in the price of x, or any other change in the production of x that can be treated as if it were a fall in price*, we call these Hicksian (gross) complementarities. In what follows we use that term complementarity in the gross rather than the net sense.

Technologies that cooperate with each other, either as subtechnologies within one main technology, or as separate stand-alone parts of some technology system, are typically net substitutes and gross complements. If the price of one technology falls, *and output is held constant*, the use of that technology will increase at the expense of most others. In what we regard as the typical case, however, this substitution effect is small enough for the income effect to dominate, making the technologies gross complements. What we have just said refers to a set of cooperating generic technologies such as the parts of an internal combustion engine or a computer and its set of peripherals. As we have already noted, however, many of these generic technologies come in competing versions, such as brands of spark plugs or makes of printers. Each of these competing versions are both net and gross substitutes for each other, a fall in the price of one leading to a reduction in the use of the other competing brands.

Technological Complementarity Consider an innovation in one technology whose full benefit cannot be reaped until many of the other technologies that are linked to it are re-engineered, and the makeup of the capital goods that embody them are altered. We refer to these responses as "technological complementarities," defined as occurring whenever a technological change in one item of capital requires a redesign or reorganization of some of the other items that cooperate with it (in its internal makeup and/or in the main technology, and/or in the technology systems of which it is a part).[6] The most important point about this type of complementarity is that the effects cannot be modeled as the consequences of changes in the prices of flows of factor services found in a simple production function. All of the action is taking place in the structure of capital and the consequent changes will typically take the form of new factors of production, new products, and new production functions.

Our histiorcal cases provide many illustrations of technological complementarities that cannot be modeled as if they were the consequences of price changes. For one example, the consequences of the introduction of electric power into factories could not be modeled as a response to a change in the price of power in a production function designed to reflect the technological requirements of steam. Even a zero price of steam power would not have led to the radical redesign of the plant which was the major source of efficiency gain under electricity (Schurr 1990; David 1991). This redesign depended on the introduction of the unit drive which attached an efficient power delivery system to each machine, something that was impossible under steam.

For a second example, the massive set of adjustments in existing and new capital structures that Fordist mass production brought about could not be modeled as resulting from a fall in the price of noninterchangeable parts would have had an impact on the automobile industry which was both quantitively smaller and qualitatively different from the revolution in the organization of production that followed from interchangeable parts.

We now note that having extensive Hicksian and technological complementarities is a necessary condition for a technology to be a GPT.

---

6. Note that neither of the distinctions complements/substitutes and net/gross can be made in the case of technological complementarities. Let technology A change. Technology B is complementary with A if a change in its specifications is now required. The complement/substitute distinction turns on being able to sign the induced change in B, which we can do when the reaction is in the quantity of output but cannot do when the reaction is a redesign of the technology. The net/gross distinction depends on being able to hold output constant, which we can do if process technologies are changing but cannot do when induced changes occur in product technologies.

Ceteris paribus, the more pervasive a technology, the more of both complementarities it is likely to have with other technologies. Because GPTs provide materials, power, transport, and ICT inputs that enter into virtually all production, and because they typically lie at the centre of large technology systems, they are vertically and horizontally linked to many other technologies. For this reason innovations in GPTs will typically induce major structural changes in many, sometimes even the great majority of, other technologies.

But having these complementarities is not sufficient because many other technologies also have them. For example, the modern shipping container is a single-purpose technology that revolutionized cargo handling and had many complementarities, causing adjustments in size of ships, layout and location of ports, handling facilities, labor skills, the design of trucks and railcars, and international location of production.

Necessary and Sufficient Conditions for a GPT
Virtually every new technology has some of the four necessary conditions for GPTs that we identified above. Early television had great scope for improvement but was limited in its variety of uses and complementarities; lightbulbs have a wide range of uses but for only one purpose, light; belts have a wide variety of uses but not a wide range of use throughout the economy; the innovation of shipping containers caused many technological complementarities as products and processes in the shipping and related industries were redesigned to accommodate them, but they have a limited variety of uses. It follows therefore that none of the conditions outlined above can be individually sufficient to identify a GPT. In order to identify GPTs by their technological characteristics, we look for technologies that have all of the four characteristics.

Definition A GPT is a technology that initially has much scope for improvement and eventually comes to be widely used, to have many uses, and to have many Hicksian and technological complementarities.

2.6 Is the Concept Useful?

Scales that measure scope for improvement, range of use, number of uses, and extent of complementarities would be densely inhabited with technologies throughtout. Thus there is no obvious break between those technologies that are just above the bottom end of any cutoff that we use to define a GPT and those that are just below it. Nonetheless, the concept of a GPT as a technology at the extreme end of all of the scales can be useful for building theories about technological change. After all, it is a common and helpful procedure to take a set of variables distributed over some scale, divide that scale into several intervals, and define a typical member of each interval.

To check on the value of our definition, we apply it to a number of cases to see if there are technologies that are inappropriately excluded or included.

2.6.1 Further GPTs

In this section we mention technologies that our definition admits as GPTs.

Lasers
Originally we did not have lasers on our list of potential GPTs. But, on close examination, they seem to satisfy our necessary and sufficient conditions. They began as a scientific curiosity with few, if any, obvious commercial applications. Today they are widely used in a growing number of disparate applications such as checking out goods at cash desks, playing CD recordings, printing hard copy, transmitting information over long distances, and substituting for a surgeon's knife. The range of applications may not yet be as wide as the other technologies on our list, but it is growing. If it continues to grow at its present rate, it will become one of the most important GPTs of the next century. Its cuts across several technology classes. Among other things it is a power delivery system, it is used as an ICT in many applications, and it is used as a cutting tool in many surgical and manufacturing operations.

Internet
In chapter 6 of this volume, Richard Harris deals with the Internet, a technology that is clearly a GPT by our definition: It is used widely and for many different purposes, it has many Hicksian and technological complementarities, and it has tremendous scope for improvement. Yet we have not listed it as a separate GPT because it is a subtechnology of what we call the modern computer-based information and communication GPT. This observation reenforces our general point about the fractal-like structure of technologies because a main GPT can give rise to derivative technologies that are GPTs in their own right. For some purposes, as in Harris's chapter, it is useful to separate out these GPTs. For other pur-

poses, as in our two chapters, it is useful to treat the entire system of technologies that derive from, and depend on, the electronic computer as a single GPT.

Organizational Technologies

At first we excluded organizational innovations, but when challenged, we found it impossible to maintain this exclusion. Our definition of a process technology as the specification of how to produce a given product, must cover more than just the machines that are employed, including as well the micro details of plant layout and other aspects of the organization of plants and firms. This raises the question of organizational GPTs, and we fine at least three in the modern history of industrial nations.

Factory System The first industrial revolution's factory system meets our criteria. It was a response to particular innovations in capital goods. The automated textile machinery was first used in cottages, in a structure suitable for hand weaving. Gradually, production was shifted to factories in which water power fairly rapidly replaced human power to drive the machines. The factories required different organizations at all levels, including finance, and different amounts of human capital. Factories led to the decline of villages and the growth of industrial cities. The largest gains in productivity were postponed until, in the early nineteenth century, steam power replaced water power in many factories, leading to efficiency-increasing redesigns of many of the automated machines. Starting in textiles, the factory system slowly spread to include virtually all of manufacturing by the end of the century.

Mass Production This system largely eliminated craft production, which uses highly skilled workers and simple but flexible tools to make exactly what the consumer wants. Mass production uses narrowly skilled professionals to design products made by unskilled or semiskilled workers tending expensive single-purpose machines, which turn out standardized products in high volume.[7] It spread from the auto industry to cover most American manufacturing, making it the dominant method of organizing American (and Canadian) manufacturing in the mid-twentieth century. We do not say much more about it here because many of the main points have already been given in the section the automobile as a GPT.

---

7. The definitions come from Woomack, Jones, and Roos (1990).

Flexible Manufacturing Unlike the factory system, lean production, flexible manufacturing, or Toyotiasm as it is variously called, was not driven by a major change in the technology of capital goods. Rather it was driven by the need of Japanese automobile makers to produce competitively at a scale much less than was possible for U.S. firms. The principles of lean production spread from the shop floor to every phase of the firm's operations, from design, to marketing, to customer follow-up (Woomack 1990; Zuboff 1984). Working in flexible work groups rather than tending one dedicated machine required a different organization of the work force, different union practices, and different human capital. Different channels of feedback from production workers to supervisors developed. The just-in-time inventory methods spread to assist industries across the whole economy and to alter the relations between assembly firms and their suppliers. The payoff was so large that Japanese automobile firms penetrated the North American and European markets in a very big way, and without the ensuing government intervention, one or more indigenous firms would probably have been eliminated. After a few very expensive false starts, North American firms have adopted the Japanese methods, first in automobile manufacture and slowly across a wide range of other products. European protectionism has kept the Japanese challenge more peripheral, and so European firms have not been forced into anything like the same amount of productivity-increasing changes as have occurred in North America.

A Caveat These three organizational innovations clearly meet our criteria for a GPT, and their effects were as large as many of the other more conventional GPTs that we have studied. However, when we expand our definition of GPTs to include organizational innovations, we are going beyond what many writers include in their concepts of technology and technological change. As we have repeatedly emphasized, classificatory systems are judged by efficiency and usefulness. In our view, there are enough similarities between innovations in capital goods and plant layouts to warrant creating another class of GPTs called organizational GPTs.

2.6.2 Near-GPTs

Some of the technologies that we have considered come close to being GPTs by virtue of sharing most of a GPT's characteristics.

Machines

Most types of machinery have such a limited variety of uses that they do not come close to qualifying as GPTs. It was suggested to us, however, that machine tools should be regarded as a GPT. Early in the nineteenth century a distinct industry grew up to produce power-driven tools for use in manufacturing. Tools were typically developed for single purposes, and a few were then found capable of being adapted for multiple uses. For example, the turret lathe, which allowed a metal part to be subjected to a series of operations without being repositioned, was originally developed for the production of pistols. It was then progressively modified and adapted to produce key parts for sewing machines, watches, typewriters, locomotives, bicycles, and automobiles. Another important machine tool, the milling machine, was eventually used in the production of tools, cutlery, locks, arms, sewing machines, textiles machinery, printing presses, scientific instruments, and locomotives (Rosenberg 1976).

Nonetheless, we rule out machine tools because their range of use is restricted to manufactuing. From the point of view of the manufacturing industry a few key machine tools qualify as GPTs; from the point of view of the economy as a whole, they do not quite fulfill our criteria of widespread use.

Single-Purpose Technologies

Another set of near-GPTs has major complementarities, and a wide range of use across the economy, but lacks the characteristic of having many different uses. As well as the modern examples we gave earlier, such as lightbulbs and shipping containers, in simpler societies many agricultural inventions also came into this class. From the time of the neolithic agricultural revolution until the nineteenth century, agriculture accounted for most of the national income in most countries. Any major innovation in agriculture therefore was widely used and had major impacts across most of the economy. The original domestication of plants, animal fertilizer, selective breeding and hybridization, the light plough, the heavy plough, crop rotation in the three-field system, and the modern green revolution have had tremendous and widespread impacts, although they are all single-purpose technologies. As society became more complex, however, the proportion of output accounted for by agriculture fell, diminishing the chances that any technological advance in agriculture would qualify as being widely used across most of the economy, and hence capable of exerting the type of economywide effects in which we are interested.

2.6.3 Uncertainties

Viewed in retrospect, the evolutionary path of a fully developed GPT has the appearance of inevitability. When the technology is in its infancy, however, an observer looking into the future cannot conceivably know if it will turn out to be a modest advance operating over a limited range, a GPT, or anything in between. For example, had the practical uses of electricity been developed 75 to 100 years earlier, Watt's low-pressure steam engine would not have developed into a GPT. Even when a technology has become established, it spreads in ways that are hard to predict (partly because emerging technological complementarities are hard to anticipate). Electricity, lasers, and steam engines provide examples of GPTs with applications that could not have been foreseen early in their evolution (Rosenberg 1996).

At this point one may wonder if the concept of a GPT is useful. If we cannot identify GPTs at their time of birth, and if their development trajectories are subject to such uncertainty, how can we theorize successfully about them? It is important to note, however, that uncertainty is present in virtually all technological developments. If uncertainty makes it impossible to theorize about GPTs, then we cannot theorize about *any* major technological change for the same reason. Since technological change is the main source of long-term economic growth, this would rule such growth out of the realm of economic analysis. Rather than take this defeatist attitude, it seems better to try to develop useful theories that can accommodate the inherent uncertainties.

Theorizing about Uncertainty

Since technological change is a path dependent, evolutionary process taking place under conditions of uncertainty, it may not be subject to the kind of testable theories as are other economic processes. Repeated experiments may not be possible, and all relevant variables (including tastes and technology) may be endogenous. Nonetheless, as with biological evolutionary processes, empirically relevant theories can be developed that uncover the system's laws of motion, even if those laws produce neither a unique stable equilibrium, nor a process that would necessarily repeat itself if it could be replayed.

Risk Analysis

Many do not agree with the above diagnosis and continue to apply maximizing theories to technological change in general and GPTs in particular.

What Requires Explanation?   49

The theoretical papers in this volume all do so. If technological uncertainty can, for some predictive purposes, be treated as if it were merely a risky situation, then such theories will be useful. However, the existence of uncertainty is hard to deny and is accepted by virtually everyone who has studied technological change in detail. What can only be decided by experiment is how far the classical techniques of risk analysis can go in studying these situations.

Limited Predictability

Since our definition of a GPT is built on technological characteristics that are time dependent, the ability to identify a GPT must also be time dependent. Technological history is full of examples both of technologies that came to very little after their potential was widely acclaimed in the early stages of their development and of technologies that were thought to be of limited use but which developed into fully fledged GPTs over subsequent decades. We have, however, set down a definition in terms of technological characteristics and listed five categories of technologies that may aid in their identification. Do these give us any degree of predictability?

Identification of Some Potential GPTs   Some technologies can be identified early on as having the potential to develop all of the characteristics of a GPT. For example, if one is told that a technology will provide very low-cost power that can be delivered anywhere in the world, one can say that this technology, whatever its engineering characteristics, has the potential to become a GPT. Although no one can be sure how such technologies will in fact evolve, they should be watched closely. While electricity and nuclear power were both in this class early in their development, electricity clearly fits our definition of a GPT, while nuclear power does not (at least *not yet*).

Impossible to Identify All Potential GPTs   There will always be new technologies that look limited to all observers but that subsequently develop into GPTs because they evolve new applications and new functions that were unsuspected at the outset. So, although we may with reasonable confidence put some new technologies into the class of potential GPTs, we cannot with equal confidence assert that the remainder have no promise of developing into GPTs.

Identification of GPTs along the Way   At some point before a GPT's full impact is felt, it will become clear that the technology is developing into GPT. For example, long before its full potential had been exploited (which is still in the future), it became apparent to many observers that electronic computers were on their way to becoming a pervasive GPT. It can be useful to identify a GPT even several decades after it has begun its evolution. This will help policy makers to understand the technological revolutions and the structural adjustments that typically accompany GPTs.

Prediction of General but not Precise Path of Development Even though in some cases there is a good chance of classifying technologies that do and do not have the potential to evolve into GPTs, uncertainty cannot be avoided in predicting the path of development. For example, electricity was recognized as a technological marvel with many possible applications long before many of its current applications were known. The process of determining the specific uses is, of course, carried out by profit-maximizing firms seeking to exploit the new GPT. Even though it is difficult to enumerate all the specific uses and complementarities that will eventually be associated with a newly identified GPT, it *is* possible to predict some specific developments and that, qualitatively, the set will be large. Although the full potential of the computer, laser, satellite, and other related ICTs were not clear at the outset, some specific developments, such as networking, were predicted well in advance, and it has been clear for some time that we are living through a profound transformation associated with these GPTs.

New GPT's Predicted? After all this analysis, can we identify future GPTs that are emerging on the current technological horizon? As we have suggested, a consideration of the characteristics of new technologies often allows us to identify potential GPTs. A consideration of their current evolutionary trajectory then allows us to assess whether or are not they are currently fulfilling their potential. Here are a few cases in point.

Biotechnology is an obvious future GPT, although it is not yet widely used in producing economic value. Many diverse possible uses have already been established and more are being discovered at a rapid pace. Many of their practical applications, however, await further reductions in costs and an assessment of their side effects. What has happened so far makes us confident that biotechnology is an emerging GPT.

Nuclear power is a different case. Low-cost, nonpolluting fusion would probably rate as an important GPT. Its effects would not be as pervasive as some other GPTs because it would likely displace the applications of existing hydroelectric generated power, at least in the first instance. Its main immediate advantage would be to free power production facilities from the constraint of being near fast-moving water or coal supplies. Thus it could affect the pattern of comparative advantage and the relative prices of many products. Low-cost fusion in micro generators would have a prodigious number of applications that would take fusion from being an important technology in electricity generation to being a GPT in its own right. Current uncertainties are such that we have no idea if low-cost, trouble-free nuclear power will be developed or if alternatives such as solar or geothermal power will dominate it.

Superconductivity is another possibility. Its potential has been celebrated for decades, so far without major applications. No one knows if the technical problems will be overcome to make it a major technological revolution. Since we can imagine several competing technologies that might make the transfer of power on a large scale unnecessary, such as small-scale solar power plants, we have no confidence in predictions that it will become a GPT within the next century.

Nanotechnology, electronic machines smaller than human blood cells, is another possible GPT of the future with a large number of potentially valuable applications. These include allowing noninvasive surgical techniques, improved tolerances in material development, and drastically lowering the cost of producing integrated circuits. The potential for a new GPT is clear, while the evolutionary path is uncertain.

2.7 Conclusion

Economic historians have long argued whether technological change is always a process of continuous small incremental changes or is episodically punctuated by large qualitative changes. Growth economists have had similar arguments concerning aggregate growth rates. In this chapter we review historical evidence concerning the invention of pervasive technologies that are qualitatively different from anything previously experienced. They appear from time to time, sometimes causing deep structural adjustments. However, as we discuss in chapter 8, technological discontinuities need not cause discontinuities in observed aggregate growth rates.

We have laid out the relevant set of technological characteristics that define GPTs, as well as detailing a set of awkward facts concerning their evolution. We hope that these facts go some way to setting the boundaries for further research. The definition we provide allows for the ex post identification of GPTs. It also allows ex ante identification of some technologies that, early in their development trajectories, have the potential to become GPTs, and others that are far enough along to be clearly developing into GPTs. An important challenge is to develop theories that incorporate more of the relevant characteristics of GPTs and are consistent with more of the facts that surround them.

References

Arthur, B. 1988. Competing technologies: An overview. In G. Dosi, et al., eds., *Technical Change and Economic Theory*. London: Pinter.

Billington, D. 1996. *The Innovators: The Engineers That Made America Modern*. New York: Wiley.

Blaug, M. 1980. *The Methodology of Economics*. Cambridge: Cambridge University Press.

Bresnahan, T. F., and M. Trajtenberg. 1992. General Purpose Technologies: "Engines of Growth"? NBER Working Paper 4148.

Cardwell, D. S. L. 1972. *Turning Points in Western Technology: A Study of Technology, Science and History*. New York: Science History Publications.

Chandler, A. 1990. *Scale and Scope: The Dynamics of Industrial Capitalism*. Cambridge, MA.: Belknap Press.

Cipolla, C. 1993. *Before the Industrial Revolution: European Society and Economy, 1000–1700*. New York: Routledge.

David, P. 1975. *Technical Choice, Innovation and Economic Growth*. Cambridge: Cambridge University Press.

David, P. 1991. Computer and dynamo: The modern productivity paradox in a not too distant mirror. In *Technology and Productivity: The Challenge for Economic Policy*. Paris: OECD.

Dosi, G., C. Freeman, R. Nelson, G. Silverberg, and L. Soete. 1988. *Technical Change and Economic Theory*. London: Pinter.

Drews, R. 1992. *End of the Bronze Age*. Princeton: Princeton University Press.

Dudley, L. 1991. *The Word and the Sword: How Techniques of Information and Violence Have Shaped Our World*. Cambridge, MA: Basil Blackwell.

Dudley, L. (forthcoming) Communications and Economic Growth, *European Economic Review*.

Forester, T., ed. 1988. *The Materials Revolution*. Cambridge: MIT Press.

Freeman, C., J. Clark, and L. Soete. 1982. *Unemployment and Technical Innovation: A Study of Long Waves in Economic Development*. London: Pinter.

Freeman, C., and C. Perez. 1988. Structural crisis of adjustment. In G. Dosi, et al. eds., *Technological Change and Economic Theory*. London: Pinter.

Freeman, C., and L. Soete, eds. 1987. *Technical Change and Full Employment*. New York: Basil Blackwell.

Gies, F., and J. Gies. 1994. *Cathedral Forge and Water Wheel*. New York: Harper Collins.

Gimpel, J. 1967. *The Medieval Machine: The Industrial Revolution of the Middle Ages*. New York: Holt, Rienehart and Winston.

Greenwood, J. 1997. *The Third Industrial Revolution: Technology, Productivity, and Income Inequality*. Washington: AEI Press.

Grossman, G., and E. Helpman. 1991. *Innovation and Growth in the Global Economy*. Cambridge: MIT Press.

Huff, T. 1993. *The Rize of Early Modern Science: Islam, China and the West*. Cambridge: Cambridge University Press.

Innis, H. A. 1951. *The Bias of Communication*. Toronto: University of Toronto Press.

Innis, H. A. 1972. *Empire and Communication*. Toronto: University of Toronto Press.

Kranzberg, M., and C. S. Smith. 1988. Materials in history and society. In T. Forester, ed., *The Materials Revolution*. Cambridge: MIT Press.

Landau, R., T. Tylor, and G. Wright, eds. 1996. *The Mosaic of Economic Growth*. Stanford: Stanford University Press.

Landes, D. 1969. *The Unbound Prometheus*. Cambridge: Cambridge University Press.

Lipsey, R. G., and C. Bekar. 1994. A structuralist view of technical change and economic growth. In *Bell Canada Papers on Economic and Public Policy, vol. 3*. Proceedings of the Bell Canada Conference at Queen's University, Kingston: John Deutsch Institute.

Mokyr, J. 1990. *The Lever of Riches: Technology Creativity and Economic Progress*. Oxford: Oxford University Press.

Nelson, R. 1995. The agenda for growth theory: A different point of view. IIASA Working Paper

Rosenberg, N. 1976. *Perspectives on Technology*. Cambridge: Cambridge University Press.

Rosenberg, N. 1994. *Exploring the Black Box: Technology, Economics, and History*. Cambridge: Cambridge University Press.

Rosenberg, N. 1982. *Inside the Black Box: Technology and Economics*. Cambridge: Cambridge University Press.

Rosenberg, N. 1996. Uncertainty and technological change. In Landau, R., Taylor, L., and Wright, G., eds., *The Mosaic of Economic Growth*. Stanford, Stanford University Press.

Schmandt-Besserat, D. 1992. *Before Writing*. Austin: University of Texas Press.

Schmookler, J. 1965. Catastrophe and utilitarianism in the development of basic science. In R. A. Tybout, ed., *Economics of Research and Development*. Columbus: Ohio State University Press.

Schmookler, J. 1966. *Invention and Economic Growth*. Cambridge: Harvard University Press.

Schmookler, J. 1972. *Patent, Invention, and Economic Change: Data and Selected Essays*. Z. Griliches and L. Hurwicz, eds., Cambridge: Harvard University Press.

Schurr, S., et al. 1990. *Electricity in the American Economy*. New York: Greenwood Press.

Von Tunzelmann, G. N. 1978. *Steam Power and British Industrialization to 1860*. New York: Clarendon Press.

Von Tunzelmann, G. N. 1995. *Technology and Industrial Progress: The Foundations of Economic Growth*. Brookfield, VT: E. Elgar.

Woomack, J. P., D. J. Jones, and D. Roos. 1990. *The Machine that Changed the World*. New York: Rawson Associates.

Zuboff, S. 1984. *In the Age of the Smart Machine: The Future of Work and Power*. New York: Basic Books.

3 A Time to Sow and a Time to Reap: Growth Based on General Purpose Technologies

Elhanan Helpman and
Manuel Trajtenberg

3.1 Introduction

In any given "era" there typically exists a handful of technologies that plays a far-reaching role in widely fostering technical change and thereby bringing about sustained and pervasive productivity gains. The steam engine during the first industrial revolution, electricity in the early part of this century, and microelectronics in the past two decades are widely thought to have had such an effect. Bresnahan and Trajtenberg (1995) refer to them as general purpose technologies (GPTs hereafter).

GPTs are characterized by the following features: (1) They are extremely pervasive; that is, they are used as inputs by a wide range of sectors in the economy. (2) Their potential for continual technical advances manifests itself ex post as sustained improvements in performance. (3) Complementarities with their user sectors arise in manufacturing or in the R&D technology.[1]

These features provide a mechanism by which the GPTs act as "engines of growth": As a better GPT becomes available, it gets adopted by an increasing number of user sectors and it fosters complementary advances that raise the attractiveness of its adoption. For both reasons the demand for the GPT increases, inducing further technical progress in the GPTs, which prompts in turn a new round of advances downstream, and so forth. As the use of a GPT spreads throughout the economy, its effects become significant at the aggregate level, thus affecting overall growth.

Helpman: Harvard University, Tel Aviv University, and CIAR. Trajtenberg: Tel Aviv University and CIAR. We thank Yossi Hadar for research assistance, and George Akerlof, Gene Grossman, Peter Howitt, Richard Lipsey, Paul Romer, and Bart Verspagen for comments on an earlier draft. Helpman also thanks the NSF for financial support.

1. Lipsey, Bekar, and Carlaw discuss the nature of GPTs in more detail in chapter 2; see their chapter for additional characteristics.

We study in this chapter the economywide dynamics that a GPT can generate. For this purpose we embody the notion of GPTs, as developed in Bresnahan and Trajtenberg (1995), into a growth model à la Romer (1990) and Grossman and Helpman (1991, ch. 3). However, since a full-fledged general equilibrium model of GPTs would be exceedingly complex and intractable, we analyze a scaled-back version in which advances in the GPTs are exogenous. As a result we ignore the feedback from user sectors to the GPTs.

We refer to each GPT as prompting the development of "compatible components." Of course these can be any sort of inputs or, more generally, complementary investments of any kind. It may help to visualize this by thinking of computers, for example, and the components as compatible software packages, or even successive generations of integrated circuits, and the components being the other parts of the appliances and instruments that incorporate those circuits.

We present the basic GPT-based growth model in the next section. In section 3.3 we analyze the long-run dynamics, in the form of repetitive cycles, that result from the arrival of new GPTs within fixed time intervals. The consequent behavior over time of GDP, total factor productivity, real wages, and factor shares are described in section 3.4. A number of important results emerge there. A typical cycle contains two distinct phases. During the first phase, output and productivity experience negative growth, the real wage rate stagnates, and the share of profits in GDP declines. The benefits from a more advanced GPT manifest themselves during the second phase, after enough complementary inputs have been developed for it. During this later phase there is a spell of growth, with rising output, real wages, and profits. Over the entire cycle the economy grows at the rate determined by the rate of advance in the GPT itself.

The growth path thus obtained in the model seems to correspond to the historical record of productivity growth following the introduction of electricity a century ago, and it may likewise resemble the economywide impact of computerization of the last few decades.

In section 3.5 we present two modifications that shed light on additional issues. First, we show how to extend the model to a multiple sector economy in order to examine the role of diffusion in the growth process. In this case the acceleration of growth in the second phase is driven both by the development of complementary inputs and by the gradual diffusion of the new GPT throughout the economy.[2] Second, we examine the

---

2. In this chapter the same intermediate inputs are used by all sectors. We modify this assumption in chapter 4. Note that each sector needs to develop its own sector-specific intermediates in order to implement a new GPT.

role of skilled and unskilled workers, where R&D is skill intensive relative to manufacturing. An economy with two types of labor may still follow the aggregate dynamics that were previously described, except that now output does not need to decline over time following the arrival of a new general purpose technology. Moreover we show that during the first phase of slow growth real wages of unskilled workers stagnate, while real wages of skilled workers increase. On the other hand, in the second phase of accelerated growth, unskilled workers make real income gains, while skilled workers may experience real gains in the early part of the phase and a loss at the end of the entire cycle. We provide concluding remarks in the closing section.

3.2 Building Blocks

Suppose that a final good is produced with the aid of a general purpose technology (GPT) i and an assembly of a continuum of components $x_i(j)$, $j \in [0, n_i]$, that have to be compatible with the particular GPT in use, where n_i denotes the number (measure) of available components. The production function is given by

$$Q_i = \lambda^i D_i, \qquad \lambda > 1, \tag{1}$$

where λ^i stands for the productivity level of GPT i, and

$$D_i = \left[\int_0^{n_i} x_i(j)^\alpha \, dj \right]^{1/\alpha}, \qquad 0 < \alpha < 1. \tag{2}$$

The elasticity of substitution between any two components is thus $1/(1-\alpha) > 1$. We assume that GPTs become available in an ordered fashion from $i = 1$ onward. Therefore with m available GPTs aggregate output of final goods equals $Q = \sum_{i=1}^{m} Q_i = \sum_{i=1}^{m} \lambda^i D_i$.[3]

Each available component is supplied by a firm that owns the property right to the component's blueprint, and all blueprints are the results of past R&D efforts. Suppliers engage in monopolistic competition. All components can be manufactured with one unit of labor per unit output, independently of the GPTs with which they are compatible. Consequently marginal costs equal the wage rate w.

As is well known (see Grossman and Helpman 1991, ch. 3), (2) implies constant elasticity demand functions,

---

3. Chou and Oz (1993) developed a model with a similar production structure, but they have not examined the broad range of issues with which we are concerned in this chapter.

$$x_i(j) = \frac{p_i(j)^{-1/(1-\alpha)} D_i}{\left[\int_0^{n_i} p_i(j)^{-\alpha/(1-\alpha)} dj\right]^{1/\alpha}}, \quad j \in [0, n_i], \tag{3}$$

where $p_i(j)$ is the price of component j for the ith general purpose technology. Under these circumstances each component manufacturer equates marginal revenue to marginal costs, and hence all components will be equally priced according to[4]

$$p_i(j) = p = \frac{1}{\alpha} w. \tag{4}$$

It follows that all components for the ith GPT are employed in equal quantities x_i and that in equilibrium

$$D_i = n_i^{(1-\alpha)/\alpha} X_i, \tag{5}$$

where $X_i \equiv n_i x_i$ represents aggregate employment of components by users of the ith GPT. Given that a unit of labor is required for the manufacturing of a unit of components, however, X_i also represents labor employment in the manufacturing of components for the ith general purpose technology.

It is clear from (1)–(5) that total labor input per unit of final output for the ith GPT is $b_i \equiv X_i/Q_i = \lambda^{-i} n_i^{-(1-\alpha)/\alpha}$. This labor input is lower for GPT i the more components are available for i. Evidently a general purpose technology is more valuable, the more compatible components have been developed for its use.

Competitive suppliers of the final output minimize unit manufacturing costs of Q. Therefore they choose to manufacture with those general purpose technologies whose productivity level λ^i in combination with the number of available components n_i yield the lowest unit costs. Thus the price of final output equals[5]

---

4. The manufacturer of $x_i(j)$ maximizes profits. Namely he chooses $p_i(j)$ to maximize $[p_i(j) - w]x_i(j)$ subject to (3). Since his input is of measure zero relative to the total employment of inputs by manufacturers of the final output, he realizes that D_i is unaffected by his choice of price. Therefore the solution to his problem is given by (4).

5. A typical producer of final output chooses from among the available GPTs, and from among the available complementary inputs for each one of them so as to minimize his unit manufacturing costs. Namely for an available GPT i he chooses inputs to obtain the lowest unit cost

$$c_i = \min_{\{x_i(j)\}} \left\{ \int_0^{n_i} p_i(j) x_i(j) \, dj; \text{ s.t. } \lambda^i \left[\int_0^{n_i} x_i(j)^\alpha \, dj\right]^{1/\alpha} \geq 1 \right\}$$

and chooses to work with the GPT that yields the lowest c_i. In view of (4) this yields (6).

$$p_Q = \frac{1}{\alpha} wb, \quad \text{where } b = \min_{1 \leq i \leq m} b_i, \ b_i \equiv \lambda^{-i} n_i^{-(1-\alpha)/\alpha}. \tag{6}$$

A technology whose unit costs exceed wb/α is not used, which implies that manufacturers of components that are compatible with this technology have no sales and make no profits.

Suppose that at first there is just one GPT, $m = 1$, and that suitable complementary inputs for it have been developed. Then a second generation GPT appears. However, in order for the second generation GPT to be used in production, appropriate complementary inputs have to be developed (the previous inputs are not compatible). The switch to the second generation GPT will occur only after "enough" such inputs have been developed, namely only after $n_2/n_1 > 1/\lambda^{\alpha/(1-\alpha)}$ (see (6)). In general, there will be a switch from the ith GPT to the $(i+1)$th as soon as

$$n_{i+1} > \eta n_i, \quad 0 < \eta \equiv 1/\lambda^{\alpha/(1-\alpha)} < 1. \tag{7}$$

Now, if Q_i units of final output are produced with the ith general purpose technology, then its users employ $x_i = b_i Q_i / n_i$ units of each one of the n_i components. In view of the pricing equation (4), this implies that each component yields a profit stream

$$\pi_i = \frac{(1-\alpha) w b_i Q_i}{\alpha n_i}. \tag{8}$$

These profits equal zero whenever $b_i > b$ (see (6)), that is, whenever GPT i requires labor input per unit output that is not the lowest.[6]

Next suppose that each firm possesses indefinite monopoly power in the supply of its component. Then the value of a firm equals the present value of its profits,

$$v_i(t) = \int_t^\infty e^{-R(t,\tau)} \pi_i(\tau) \, d\tau, \tag{9}$$

where $R(t, \tau) = \int_t^\tau r(z) \, dz$ stands for the discount factor from time τ to t and r is the interest rate. Differentiation with respect to t yields the

---

6. It is important to note that a supplier of a component for an inferior GPT i (i.e., a GTP that yields a unit manufacturing cost for final output c_i such that $c_i > \min_l c_l$) cannot sell his product even if he reduces his price down to the marginal manufacturing costs w because he takes D_i as given and $D_i = 0$ for an inferior technology. Put differently, in these circumstances an individual supplier of a single input cannot generate demand for his product by cutting price. If all suppliers of the inputs for GPT i could coordinate their pricing policy, they might be able to jointly reduce prices of the intermediate products so as to generate for them positive demand in a profit-enhancing way. But this possibility does not exist in the decentralized economy of the type we consider.

no-arbitrage condition

$$\frac{\pi_i}{v_i} + \frac{\dot{v}_i}{v_i} = r. \tag{10}$$

This condition is satisfied at each point in time for each existing component.

We assume that the R&D process rendering new components over a time interval of length dt takes the form $\dot{n}_i\, dt = (l_i/a)\, dt$, where l_i is the amount of labor devoted to the development of new components for the ith GPT. Each new component is worth v_i in present value terms. Therefore the employment of l_i workers in product development over a time interval of length dt generates the value $v_i l_i/a$ per unit time at a cost of wl_i per unit time. Net value-maximizing investors in R&D will thus abstain from product development whenever $v_i < wa$. On the other hand, $v_i > wa$ provides infinite profit opportunities, which cannot prevail in equilibrium. It follows that free entry into R&D implies the equilibrium condition

$$v_i \leq wa \quad \text{with equality whenever} \quad \dot{n}_i > 0. \tag{11}$$

Evidently, if new components are developed at all, they are developed for those GPTs that have the highest valued components. As a result, whenever it is profitable to develop components for the ith GPT, it is not profitable to develop components for older general purpose technologies.

In this type of economy labor demand arises from two sources: product development, which uses $a\dot{n}_i$ units of labor, and the manufacturing of components, which requires $b_i Q_i$. Therefore an economy with m general purpose technologies and L units of labor faces the resource constraint

$$a \sum_{i=1}^{m} \dot{n}_i + \sum_{i=1}^{m} b_i Q_i = L. \tag{12}$$

In this equation the flow of new products $\dot{n}_i$ has to be nonnegative for all i.

To complete our model, we need to specify intertemporal preferences and the appearance of new general purpose technologies. We will deal with the latter in the next section. As for preferences, we assume that at each point in time t consumers allocate consumption over time so as to maximize a logarithmic intertemporal utility function $\int_t^\infty \exp(-\rho \tau) \log C(\tau)\, d\tau$, where ρ is the subjective discount rate and $C(\tau)$ is consumption at time τ. Consumers of this type, who face an intertemporal budget constraint according to which the present value of consumption equals the present value of income plus the value of initial asset holdings, allocate consumption according to the rule $\dot{C}/C = r - \dot{p}_Q/p_Q - \rho$. And in equilibrium $C = \sum_{i=1}^{m} Q_i$. We normalize total consumer spending to equal a constant value E at each point in time, and we choose for simplicity $E = 1$. This normalization entails no loss of generality. In this event the value of output equals one at each point in time, which together with the differential equation for consumption implies that the nominal interest rate r equals the subjective discount rate. Namely

$$p_Q(t) \sum_{i=1}^{m} Q_i(t) = 1 \quad \text{and} \quad r(t) = \rho \quad \text{for all } t. \tag{13}$$

For further details about this specification see Grossman and Helpman (1991, ch. 3).

3.3 Dynamics and the Impact of New GPTs

Our analysis of the system's dynamics will focus on the wage rate and on the number of available components. This section is in three parts. First we discuss the arrivals of new general purpose technologies and the phases of typical long-run cycles. Then we characterize the motion of the wage rate and the number of components within each phase separately. Finally we specify the equilibrium links between phases and the resulting global dynamics.

3.3.1 Arrivals of New GPTs

Suppose that new GPT's arrive at predetermined time intervals of equal length $\Delta = T_i - T_{i-1}$, where T_i represents the time at which the ith general purpose technology becomes available. We will refer to each such time interval as a "cycle," and to subperiods within it as "phases" of the cycle. In what follows we focus on long-run equilibria in which each cycle looks the same, except for suitable factors of proportionality whose precise nature we explain below. We assume that the development of components for a particular GPT cannot start before the GPT is actually introduced.[7] There are two possible scenarios in this case, depending on

---

7. Clearly that is not always the case in reality (e.g., some software for a new operating system is developed before the launching of the latter). In this case, however, the developers of components need to know enough about the upcoming GPT in order to start their own R&D early on. This raises important issues regarding the flows of information between the GPT and the user sectors, that may impinge on the growth process (see Bresnahan and Trajtenberg 1995).

how long Δ is: (1) that a moment before T_{i+1} components for GPT i are still being developed, and hence the appearance of the new GPT brings to an abrupt end those developments; and (2) that the development of the components compatible with the ith GPT has ceased before T_{i+1}, and hence after a hiatus when there is no product development, R&D for the new components starts at T_{i+1}.

In case 1 the sequence of events is as follows: Recall from (7) that production with GPT $i+1$ becomes profitable as soon as $n_{i+1} \geq \eta n_i$. We denote by Δ_1 the length of time that elapses from the arrival of a new GPT up to the point in time at which it becomes profitable to switch to the new GPT in the production of components. During this time interval the *old* GPT is used to manufacture final output, while R&D is used to develop components for the *new* GPT. We call this initial period "phase 1." In phase 2, which is of length $\Delta_2 \equiv \Delta - \Delta_1$, the old GPT is superseded by the new one in production while the development of components continues for the new GPT. Since we assume that Δ and λ (and hence η) are constant, so will be Δ_1 and Δ_2 in a stationary long-run equilibrium.

Unlike case 1, in case 2 the development of components for the ith GPT ceases at $T_i + \Delta_n < T_{i+1}$ rather than at T_{i+1}, while other features remain the same as in case 1. It follows that case 2 has in addition to phases 1 and 2 a third phase in the time interval $[T_i + \Delta_n, T_{i+1})$, during which production takes place with the new GPT and no R&D takes place whatsoever.

3.3.2 Phase 1

Consider the time interval $[T_i, T_i + \Delta_1)$ during which producers of final output employ the $(i-1)$th general purpose technology, but innovators invest resources in order to develop components for the ith GPT. During this phase the number of components compatible with the old GPT remains constant at the level $n_{i-1}(T_i)$, and profits for the upstarts equal zero. In this event the no-arbitrage condition (10), together with the free-entry condition (11) and the normalization (13), implies that

$$\frac{\dot{w}}{w} = \rho \quad \text{for } t \in [T_i, T_i + \Delta_1).$$

During this phase innovators make no profits but nevertheless engage in innovation because they expect profits in the future, and they are indifferent as to when to innovate, since R&D costs are rising at the rate of interest rate. As a result the capital value of R&D costs is the same at each point in time. In addition (6), (12), and (13) imply that

A Time to Sow and a Time to Reap

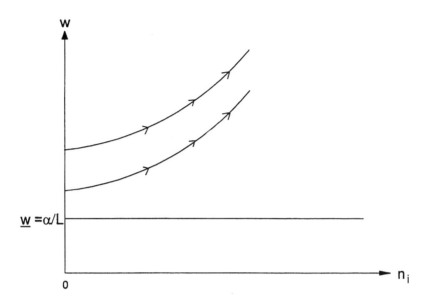

Figure 3.1
Wage dynamics in phase 1

$$\dot{n}_i = \frac{1}{a}\left(L - \frac{\alpha}{w}\right) \quad \text{for } w \geq \frac{\alpha}{L}, \ n_i(T_i) = 0, \ t \in [T_i, T_i + \Delta_1].$$

The condition on the wage rate ensures nonnegative employment in R&D. Given an initial wage rate $w(T_i)$, these differential equations yield the following solutions for the evolution of the wage rate and the number of products during phase 1:

$$w(t) = w(T_i)e^{\rho(t-T_i)} \quad \text{for } t \in [T_i, T_i + \Delta_1], \tag{14}$$

$$n_i(t) = \frac{L}{a}(t - T_i) - \frac{\alpha}{\rho a w(T_i)}[1 - e^{-\rho(t-T_i)}] \quad \text{for } t \in [T_i, T_i + \Delta_1]. \tag{15}$$

Figure 3.1 describes two feasible trajectories of system (14)–(15) (the higher trajectory begins with a higher initial wage rate). As can be seen, both the wage rate and the number of products rise over time along these trajectories.

3.3.3 Phase 2

Now suppose that we are in the time interval $[T_i + \Delta_1, T_i + \Delta_e)$, where $\Delta_e = \Delta$ if case 1 applies and $\Delta_e < \Delta$ if case 2 applies. In either case during

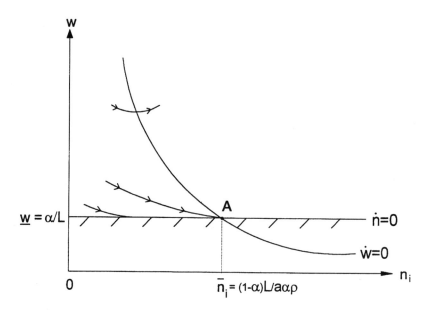

Figure 3.2
Wage dynamics in the last phase

this phase GPT i has the lowest unit labor requirement coefficient b_i, and all manufacturers of final output employ this technology. In addition it is profitable to keep investing in the development of new components for this general purpose technology, and no R&D takes place for components of other GPTs. In this event (11) is satisfied with equality, which together with (6), (8), (10), and (13), implies that

$$\dot{w} = \rho w - \frac{1-\alpha}{a n_i} \quad \text{for } t \in [T_i + \Delta_1, T_i + \Delta_e). \tag{16}$$

This differential equation for the wage rate holds as long as the ith GPT remains the best practice technology, and new components can profitably be developed for its use. Next observe that as long as i remains the lowest cost GPT, (6), (12), and (13) imply, as before, that

$$\dot{n}_i = \frac{1}{a}\left(L - \frac{\alpha}{w}\right) \quad \text{for } w \geq \frac{\alpha}{L},\ t \in [T_i + \Delta_1, T_i + \Delta_e). \tag{17}$$

Equations (16) and (17) describe an autonomous system of differential equations whose motion we depict in figure 3.2. The hyperbola $\dot{w} = 0$ describes the rest points of (16) while the horizontal line $\dot{n} = 0$ describes the rest points of (17). The shaded area below the horizontal line identifies a region in which investment in R&D is negative. Therefore this region is not feasible. Three arrowed curves describe dynamic trajectories. The intermediate trajectory is a saddle path that converges to the stationary point A. If no new GPTs are expected to become available, then this saddle path is the unique equilibrium trajectory that satisfies the valuation equation (9) and the intertemporal budget constraint (see Grossman and Helpman 1991, ch. 3). If, on the other hand, new GPTs are expected to appear in the future, as described in section 3.3.1, then the equilibrium trajectory depends on the length of a cycle and on how much better a new technology is relative to the old.

3.3.4 Phase 3

In phase 3 (applicable to case 2 only) there is no R&D, all resources are employed in manufacturing, and all manufacturers of final output employ the newest general purpose technology. In this event the number of components is constant, and so is the wage rate.

3.3.5 Global Dynamics

We proceed now to examine separately each of the two cases identified in section 3.3.1. The analysis throughout is predicated on the assumption of perfect foresight.[8]

Case 1
Recall that in this case a cycle consists of two phases. In the first phase final output is manufactured with the old general purpose technology and innovators develop components for the new GPT. The evolution of the wage rate and the number of components of the new GPT follow the pattern that we discussed in section 3.3.2, namely both rise over time. In the second phase, after sufficiently many complementary inputs have been developed for the new general purpose technology, manufacturers of final output switch to the new GPT, while innovators continue to develop

---

8. There also exists a degenerate equilibrium with no product development for new general purpose technologies. Suppose that when a new GPT i arrives, every potential innovator expects that no one will invest in R&D in order to develop components for its use; that is, everyone expects $n_i(t) = 0$ for all time periods. As a result it does not pay to invest in the development of a single component because the new GPT will never be used in manufacturing. In this event the pessimistic expectations are self-fulfilling, and no new GPTs are implemented. This is an expectations driven equilibrium that leads to stagnation, but it is a decentralized equilibrium nevertheless. We do not discuss these types of equilibria in what follows.

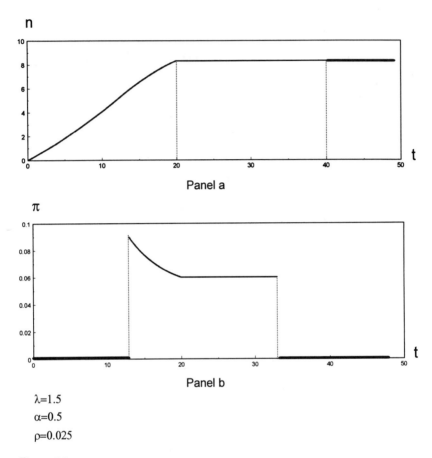

Figure 3.3
Number of products and profits

components for the new GPT. It follows that in the second phase, when the new GPT dominates both production and product development, the dynamics of the wage rate and the number of components (for the new GPT) follow the pattern described in section 3.3.3. It leaves open the question which of the many possible trajectories prevail in equilibrium. To answer this question, we need to identify appropriate links between the two phases, to which we turn next.

Consider a typical cycle, one that starts at T_i and ends at T_{i+1}. At T_i the ith general purpose technology becomes available. At that point in time the number of components that have been developed for the $(i-1)$th GPT has reached its peak $n_{i-1}(T_i)$, and innovators switch their efforts to develop complementary inputs for the new i^{th} GPT. The evolution of $n_i(t)$

A Time to Sow and a Time to Reap 67

is described in panel (a) of figure 3.3. Starting from zero at T_i, the number of components for the new GPT grows over time until a still newer general purpose technology appears at T_{i+1}. From that point on the number of components n_i remains constant at the level $n_i(T_i)$, while innovators switch to develop components for the $(i+1)$th GPT.

Panel b of figure 3.3 describes the evolution of profits. Profits π_i equal zero as long as manufacturers of final output employ the $(i-1)$th GPT. This changes at $T_i + \Delta_1$, at which point the number of components available for the ith GPT, $n_i(T_i + \Delta_1)$, makes manufacturers of final output just indifferent between the old and the new general purpose technology. From (7) this implies that

$$n_i(T_i + \Delta_1) = \eta n_{i-1}(T_i). \tag{18}$$

From this point on profits become positive for suppliers of components for the ith GPT. Since profits are (see (6), (8), and (13))

$$\pi_i(t) = \frac{1-\alpha}{n_i(t)} \quad \text{for } t \in [T_i + \Delta_1, T_{i+1} + \Delta_1), \tag{19}$$

these profits fall as the number of components compatible with the new GPT rises over time. The rise of n_i during phase 2 is brought to a halt with the appearance of a still newer GPT at T_{i+1}. From this point on π_i remains constant, while innovators develop complementary inputs for the $(i+1)$th GPT. Profits drop to zero when enough inputs have been developed to induce manufacturers of final output to abandon the ith general purpose technology and to adopt the $(i+1)$th. They remain zero thereafter.

Consider a manufacturer of a component for the $(i-1)$th GPT. She knows that her profits will remain constant at the level $(1-\alpha)/n_{i-1}(T_i)$ over the time interval $[T_i, T_i + \Delta_1)$ and drop to zero thereafter, as described above. In this event (9), (13), and (19) imply that the value of her firm will be

$$v_{i-1}(t) \equiv \int_t^{T_i + \Delta_1} e^{-\rho(\tau-t)} \pi_{i-1}(\tau) \, d\tau$$

$$= \frac{(1-\alpha)}{\rho n_{i-1}(T_i)} [1 - e^{-\rho(T_i + \Delta_1 - t)}] \quad \text{for } t \in [T_i, T_i + \Delta_1). \tag{20}$$

Now, in order for the development of the "last" component for GPT $i-1$ at time T_i^- to have taken place, it must have been that $v_{i-1}(T_i^-) = w(T_i^-)a$ (see (11)). This "free entry" condition, together with (20) and the continuity of $v_{i-1}(t)$ that follows from (9), implies that

$$w(T_i^-) = \frac{(1-\alpha)}{p a n_{i-1}(T_i)}(1 - e^{-\rho \Delta_1}). \tag{21}$$

This equation then establishes the wage rate that prevails at the end of a cycle.

Just before and just after $T_i + \Delta_1$ (the point in time at which producers of final output switch to the ith general purpose technology) innovators invest in R&D in order to develop components for the ith GPT. In this event $v_i(t) = w(t)a$ just before and just after $T_i + \Delta_1$. Since by (9) the value of a firm is continuous in time, it follows that the wage rate is also continuous at $T_i + \Delta_1$. Equation (14) implies then that

$$w(T_i + \Delta_1) = w(T_i)e^{\rho \Delta_1}. \tag{22}$$

In addition (15) implies that

$$n_i(T_i + \Delta_1) = \frac{L}{a}\Delta_1 - \frac{\alpha}{paw(T_i)}(1 - e^{-\rho \Delta_1}). \tag{23}$$

Now we have a complete system. It consists of two sets of dynamic equations (14)–(15) and (16)–(17) for the wage rate and the number of components, the initial conditions $n_i(T_i) = 0$, and the connecting equations (18), (21), (22), and (23). The system has a long-run stationary equilibrium with cycles of constant length, which are depicted in figure 3.4 for Δ_1 constant, $w(T_i) = w(T_{i-1})$, $w(T_i^-) = w(T_{i-1}^-)$, and $n_i(T_{i+1}) = n_{i-1}(T_i)$.[9] In the first phase of each cycle the wage rate rises, and it declines in the second phase. With the appearance of a new general purpose technology the wage rate jumps upward, which makes it unprofitable to further invest in the development of components for the old GPT (the old GPT is at the margin of profitability before the appearance of the new one). If the appearance of a new GPT does not bid up wages then case 1 rather than 2 applies.

In order to compute the stationary long-run equilibrium, we seek a fixed point in (w, n) space. We begin by specifying initial values for the wage rate at the beginning of a cycle $w(T_i) = w_0$, and for the number of components at the end of a cycle $n_{i-1}(T_i) = n_{\max}$. Using these initial values, we calculate $w(T_i + \Delta_1)$, $n_i(T_i + \Delta_1)$, and Δ_1 from (18), (22), and

---

9. The stationarity conditions for the span of the first phase and the number of components are independent of normalization. On the other hand, the stationarity conditions on the wage rate result from our normalization (13). Alternatively, and independently of normalization, we could have expressed the stationarity conditions on the wage rate in terms of real rather than nominal wages.

W

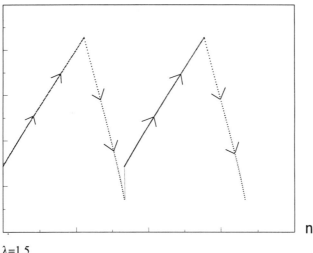

λ=1.5
α=0.5
ρ=0.025

Figure 3.4
Wages

(23). Then we apply the differential equations (16) and (17) for a time interval of length $\Delta - \Delta_1$, starting with the initial values $w(T_i + \Delta_1)$ and $n_i(T_i + \Delta_1)$, in order to compute $w(T_{i+1}^-)$ and $n_i(T_{i+1})$. If $n_i(T_{i+1}) = n_{\max}$ and $w(T_{i+1}^-)$ satisfies (21), namely if

$$w(T_{i+1}^-) = \frac{(1-\alpha)}{\rho a n_{\max}} (1 - e^{-\rho \Delta_1}),$$

then we have found a fixed point. Otherwise, we adjust the initial values and repeat the previous steps until we find a fixed point. Once a fixed point has been found, we can use the initial values $(w_0, n_{\max})$ to calculate a representative cycle of the long-run equilibrium trajectory.

Case 2
This case is similar to case 1, except that at the end of phase 2 the wage rate has to equal α/L, and it remains at this level for the rest of the cycle (this wage rate clears the labor market in the absence of employment in R&D). In this event $w(T_{i+1}^-)$ is known to equal α/L. Therefore, instead of $w(T_{i+1}^-)$, we now calculate the length of phase 2, Δ_n, which is shorter than $\Delta - \Delta_1$. This is done by running the differential equations (16)–(17) from

the initial values $w(T_i + \Delta_1)$ and $n_i(T_i + \Delta_1)$ until the wage rate drops to α/L.

3.4 Dynamics of Key Economic Aggregates

As mentioned in the introduction, our prime goal is to understand the role of GPTs and of complementary investments in economic growth. Thus we proceed now to analyze the dynamic behavior of output and of other economic aggregates.

3.4.1 GDP Growth

The real gross domestic product represents a standard measure of an economy's real output. GDP can be measured either on the output or on the input side (both lead of course to the same result). In our case GDP consists of wages plus profits on the input side. From (6) and (8) profits equal the fraction $1 - \alpha$ of the value of final output $p_Q Q$, and the latter equals one by our normalization (13). It follows that nominal GDP equals $wL + 1 - \alpha$. To calculate real GDP, we divide nominal GDP by the price of final output p_Q as defined in (6), yielding the following measure:

$$G(t) = \left[L + \frac{1-\alpha}{w(t)}\right] \alpha \lambda^k n_k(t)^{(1-\alpha)/\alpha}, \tag{24}$$

where k is the index of the general purpose technology actually used in the production of final output.[10] Observe that the term outside the square brackets on the right-hand side of (24) is continuous in time in view of the fact that each n_i is continuous in time and the GPT switching condition (7). Therefore real GDP is continuous as long as the wage rate is continuous. Recall, however, that the wage rate jumps upward with the appearance of a new general purpose technology, when innovators abandon the development of components for the old GPT and redirect their innovative efforts to the development of complementary inputs for the new GPT. As a result real GDP jumps downward at the beginning of each new cycle. It is important to note that this initial fall in GDP (which includes the output of the R&D sector) stems from the (static) allocative inefficiency caused by monopolistic competition in the manufacturing of components. The

---

10. In other words,
$$k = \begin{cases} i-1 & \text{for } t \in [T_i, T_i + \Delta_1), \\ i & \text{for } t \in [T_i + \Delta_1, T_{i+1}). \end{cases}$$

departure from marginal cost pricing there implies that the manufacturing sector is too small to begin with relative to the R&D sector (where "output" is priced according to marginal costs; i.e., $v = wa$). When a new GPT appears, the upward jump in the wage rate causes a sudden diversion of resources away from manufacturing toward R&D, further enhancing the initial distortion and bringing about the fall in GDP.[11]

Beyond the fall in output right at the start of a cycle, it is apparent from (24) that real GDP keeps declining throughout the first phase. This is due to the fact that the wage rate rises during the first phase of a cycle, while the number of components available for production of the final good (i.e., those associated with the previous GPT) remains constant. We conclude that (temporary) reductions in real output are an integral feature of the long-run equilibrium.

Actual growth begins only in the second phase, when the economy has just developed "enough" components for the new GPT and hence production of final goods switches to the newest technology. Real GDP increases throughout the second phase as the number of components for the prevailing GPT keeps rising, and as the wage rate declines. Thus it is only in phase 2 that the opportunities opened by the advent of a new GPT translate into a growth spell that continues until an even newer GPT appears. Figure 3.5 shows the simulated path of real GDP over two cycles, highlighting the sharp differences between the two phases in each cycle. Over an entire cycle, though, the average rate of growth is simply $g = \log \lambda / \Delta$.

The phenomenon sketched in figure 3.5 underlies the critical role of complementary investments in the growth process: Contrary to what happens in the context of the neoclassical growth model with exogenous technological progress, in our context technical progress in a key technology does not bring about growth in total factor productivity by itself, and certainly not right away (in our model total factor productivity growth is indistinguishable from GDP growth). Rather, the appearance of a new, more efficient GPT *induces* an initial deployment of resources into complementary investments (i.e., the development of compatible components), and it is only after there are enough such investments in place that the potential of the new technology begins to manifest itself in output

---

11. Of course, without positive markups in the manufacturing of components new ones would never be developed, and likewise, without a redeployment of resources away from the old GPT-components, the economy would never reap the benefits of the new technologies. This is then a further example of Schumpeter's "gales of creative destruction" and the consequent trade-offs between static and dynamic efficiency.

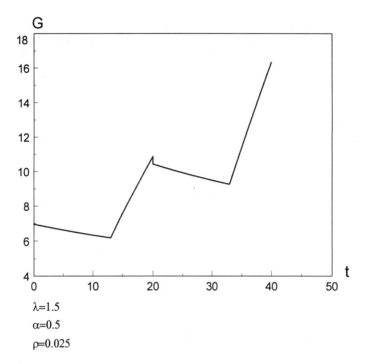

$\lambda=1.5$
$\alpha=0.5$
$\rho=0.025$

Figure 3.5
Output

and productivity gains. The economy suffers TFP (total factor productivity) and output *losses* during the first phase of a cycle, but output and total factor productivity gains re-appear with a vengeance in the second phase of each cycle. Observe, however, that the economy experiences average output and TFP growth over an entire cycle.[12]

The history of technology offers supportive evidence to this highly stylized sequence. In fact economists and other scholars have been repeatedly puzzled by the fact that new GPTs fail to deliver noticeable benefits for quite a long time after introduction, but then they "kick off" and ignite a spell of sustained growth. That was certainly the case with electricity, and that seems to be the case presently with computers and the associated "productivity puzzle." David's (1991) documentation of the introduction of electricity is of particular interest. As he has shown, substantial productivity gains appeared only about four decades after the

---

12. In Helpman and Trajtenberg (1994) we report simulation results for a wide range of parameter values. They all show a similar pattern of dynamics.

introduction of dynamos. And his figure 4a, which depicts the evolution of productivity during the diffusion of electricity in manufacturing, resembles our figure 3.5.[13] Surely there are other forces at work behind the productivity path of GPTs (e.g., plain diffusion as in Griliches 1957, or continuous improvements in the GPT itself as in Bresnahan and Trajtenberg 1995), but it seems that complementary investments play a critical role, which has been largely overlooked.[14] In this simple model those investments take the form of compatible inputs, but of course in actuality there is a wide array of different types of complementary investments, including organizational and institutional changes, both within firms and across vertically related firms.

Thus, for example, it is becoming quite clear that in order to reap the benefits from computerization, firms have to redesign the organization of work (e.g., emphasize teamwork rather than hierarchical links), decentralize decision making, and make flexibility a prime goal in planning production and product design. In contrast to the traditional view of organizations as exogenous and of organizational changes as costless, the history of technology suggests that changes in technology and changes in organization and institutions are intimately related (see Chandler 1977), and that tangible investments in such changes in response to the opportunities offered by new GPTs may be crucial for growth.

In our model the distinction between the two phases of the cycle is very pointed, and involves sharp discontinuities. This is for the most part an artifact due to the fact that we do not allow new and old GPTs to coexist, neither in production nor in R&D.[15] Such coexistence would smooth out the transition from phase 1 to phase 2, and may even reverse the negative growth of phase 1. Still, growth during the second phase would be significantly faster than in phase 1. Thus the main inference from our analysis is that even if substantially more efficient, new technologies may barely make a dent at first in actual growth, since they have to await for the development of a sufficiently large pool of complementary assets to make a significant and lasting impact. Moreover these assets

---

13. Figure 3.8, shown in section 3.6, resembles David's figure 4a even more closely.
14. See, however, Milgrom and Roberts (1990) for the importance of complementarities in manufacturing.
15. We started our work on this project with a model that allowed for overlapping GPTs, but it soon became clear that it would be exceedingly difficult to work with it because of the need to keep track in the dynamic calculations of the variables associated with all previous GPTs (primarily n_i). However, we can envision extending the model to allow for the coexistence of two contiguous GPTs at a time.

use up resources, and hence in the short run, growth may be adversely affected.

3.4.2 Real Wages, Profits, and Factor Shares

In previous sections we discussed the behavior of the wage rate over time in units that reflect the particular normalization chosen there (i.e., equating the value of final output to one). We turn now to *real* wages and profits, which measure the return to factors in units of the final good, namely w/p_Q and π/p_Q. In view of (6), the real wage rate at time t equals $\alpha \lambda^k n_k(t)^{(1-\alpha)/\alpha}$, and real profits are $(1-\alpha)\alpha \lambda^k n_k(t)^{(1-\alpha)/\alpha}/w(t)$. Notice that the continuity of n_k and the switching condition (7) imply that the real wage rate is continuous. In fact real wages remain constant during phase 1 of each cycle and then rise over the course of phase 2, as more and more components of the new GPT are being developed (hence boosting labor productivity). Real profits, on the other hand, do drop at the beginning of each cycle as a result of the upward jump in the (normalized) wage rate, and decrease further along phase 1 (recall that real GDP declines over that period and that the real wage bill remains constant). During phase 2 though, real profits rise continuously.

As to factor shares, the labor share is $wL/(1-\alpha+wL)$, and the share of profits is $(1-\alpha)/(1-\alpha+wL)$. Recalling the behavior of w over time, it is thus clear that the profit share is procyclical (as is the case with real profits), whereas the labor share is countercyclical.

3.4.3 Stock Market

The fact that the appearance of new GPTs eventually renders older ones obsolete, implies that the know-how for the manufacturing of at most two types of components can simultaneously command a positive value; those associated with the latest GPT and with the next to the last. When the economy is in the second phase of a typical cycle, only components of the best practice GPT are valuable because at that time it is known that no component of older technologies will ever be used. On the other hand, when the economy is in phase 1, then components of both the best practice GPT and of the previous one have positive value. This is so because in phase 1 the older GPT is still used in manufacturing, thereby providing a profit stream to owners of components that go with it. At the same time, owners of components that go with the newer GPT do not collect

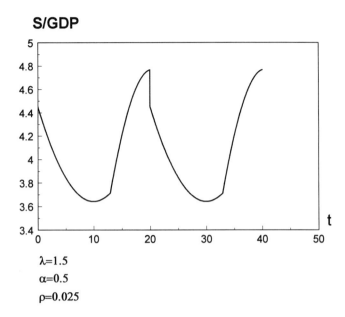

Figure 3.6
Stock market to GDP ratio

profits, though they expect profits in the future. For these reasons the technological know-hows of both types of components are valuable.

It follows that the value of the stock market at time t can be expressed as $S(t) = n_{i-1}(t)v_{i-1}(t) + n_i(t)v_i(t)$, where i is the index of the best practice GPT at time t. In case 1, on which we focus, the value of a component for the best practice technology always equals R&D costs, namely $v_i(t) = w(t)a$. As for the $(i-1)$th GPT, during phase 1 the value of each component is given by (20), while in phase 2 their values equal zero. Therefore the value of the stock market is

$$S(t) = \begin{cases} \frac{1-\alpha}{\rho}[1 - e^{-(T_i + \Delta_1 - t)}] + n_i(t)w(t)a & \text{for } t \in [T_i, T_i + \Delta_1), \\ n_i(t)w(t)a & \text{for } t \in [T_i + \Delta_1, T_{i+1}). \end{cases} \quad (25)$$

We first use this formula to simulate the stock market to GDP ratio $S/(1 - \alpha + wL)$ (see figure 3.6). Notice that the introduction of a new GPT brings not only to a sharp decline in real GDP, but to an even sharper decline in the real value of the stock market, so that the S/GDP ratio falls. This ratio continues to fall during a substantial part of phase 1, but it picks up toward the end of the phase. Thus an upturn in S/GDP

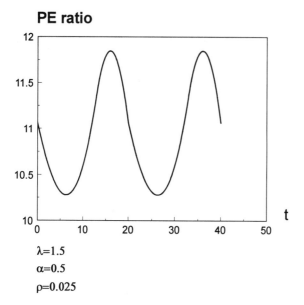

Figure 3.7
Price-earning ratio

precedes the arrival of phase 2, and thereby predicts the forthcoming upturn in productivity and output growth. In the second phase the stock market outperforms output and productivity, and the ratio of the stock market to GDP rises.

Next we consider the price-earning ratio. In phase 2, when only the best practice GPT is valued, all components with positive stock market values have the same PE ratio. But in phase 1, in which components of two GPTs are positively valued on the stock market, each type has a different PE ratio. In particular, since components of the best practice technology have no earnings but are positively valued nevertheless, their PE ratio is infinite. In either case, however, we can calculate the average PE ratio, as the value of the stock market divided by aggregate profits, $S(t)/(1-\alpha)$.

Figure 3.7 presents a plot of the price-earning ratio over two consecutive cycles. In phase 1 the PE ratio declines initially and rises subsequently, while in phase 2 it rises initially and declines subsequently. The average value of this ratio is, however, lower during the first phase, when real output and productivity decline, and higher in the second phase, when real output and productivity rise. Two points are worth making about the plot in figure 3.7. First, it shows that the price-earning ratio is more volatile than output and also more volatile than the S/GDP ratio. Second, it shows that upward trends as well as downward trends in the PE ratio can take place both during periods of economic contraction (phase 1) and during periods of expansion (phase 2).

3.5 Technological Diffusion and Relative Wages

We present in this section two modifications of our "base case" model that help address a whole set of interesting, more micro-oriented issues.[16] The first extends the model to an economy with many sectors, with the GPT having a different productivity impact on each one of them. As a result new GPTs spread gradually across them, and growth is seen to depend not only on the development of complementary inputs as in the base case but also on the rate of diffusion of the GPT. The second modification allows for the existence of skilled and unskilled labor, with R&D being relatively skill intensive. We investigate in this framework the evolution of *relative* and *real* wages over each of the phases of the cycle, and the growth of GDP. The analysis has a bearing on the recent debate concerning the observed decline of real wages of unskilled workers in industrial countries during the 1980s.

3.5.1 *Diffusion*

Suppose that there exists a continuum of final goods, indexed by z, with $z \in [0,1]$. Each z uses for manufacturing the same GPT-compatible components, as described in section 3.2, except that now we allow the productivity level of the GPT to differ across sectors. In particular, suppose that $\lambda = \lambda(z)$ is a declining function of z. That is, the productivity of a new GPT is highest for sector $z=0$, and it declines as we move to higher index sectors. Under these circumstances the price of good z equals $p_Q(z) = \alpha^{-1} w \min_{1 \leq i \leq m} [\lambda(z)^{-i} n_i^{-(1-\alpha)/\alpha}]$ (see (6)). Observe that the number of available components is the same for all goods. Thus, as the number of available components for a new GPT increases over time, a larger fraction of sectors switches to manufacture with it. In fact, given $\lambda(\cdot)$, we can derive a nondecreasing function $f(n_i/n_{i-1})$, which describes the fraction of sectors that manufacture with the newest general purpose technology i. This fraction encompasses all sectors $z \in [0, Z_i]$, where $Z_i = f(n_i/n_{i-1})$.

---

16. We refer to the model developed in previous sections as the "base case" model.

Next, consider the profits derived from a typical component, assuming that consumers allocate equal fractions of spending to all final goods. During the first phase of cycle i, profits are the same as in the base case. That is to say, they are zero for components of the ith GPT (the latest), and they equal $\pi_{i-1} = (1-\alpha)/n_{i-1}$ for components of the previous GPT. Unlike in the base case, however, at time $T_i + \Delta_1$ not everyone switches to the ith GPT simultaneously. Rather, only sector $z=0$ does so. But as more components for the ith GPT are developed over time, more and more sectors adopt the latest GPT. Thus during the second phase the profit streams are different from those that obtain in the base case. For those still manufacturing components for the older GPT, profits are $\pi_{i-1} = (1-\alpha)(1-Z_i)/n_{i-1}$, whereas in the base case nobody does so, and hence $\pi_{i-1} = 0$. For those manufacturing components for the latest GPT, profits are $\pi_i = (1-\alpha)Z_i/n_i$, whereas in the base case they were just $\pi_i = (1-\alpha)/n_i$.

Our discussion assumed that by time T_{i+1} (i.e., by the end of the ith cycle) all sectors have adopted the ith GPT, namely $Z_i(T_{i+1}) = 1$. This ensures the existence of the two phases that were central in our base case. It might happen, however, that some sectors still have not adopted the ith GPT as an even newer GPT appears. If the latter is true, then long-term equilibria may involve longer cycles, defined not by the time of appearance of new GPTs (as in the base case) but by the time when all sectors switch to the latest GPT. This may offer a richer (and perhaps more realistic) characterization of technologically driven long cycles.

Turning now to GDP growth, it can be shown that real GDP in units of aggregate consumption equals now

$$G = \left(L + \frac{1-\alpha}{w}\right)\alpha n_i^{((1-\alpha)/\alpha)Z_i} n_{i-1}^{((1-\alpha)/\alpha)(1-Z_i)} e^{\Lambda},$$

where $\Lambda = i \int_0^{Z_i} \log \lambda(z)\,dz + (i-1)\int_{Z_i}^1 \log \lambda(z)\,dz$. Comparing it to (24), we see that in the base case only additions of new components and the concomitant changes in the wage rate drive growth during phase 2. In the present case there is an additional force contributing to productivity growth, and that is the rate of diffusion of the latest GPT in the manufacturing sector, namely the rise in Z_i.

3.5.2 Relative Wages

Suppose now that the labor force is not uniform but rather that there are skilled workers that are suitable primarily for R&D and unskilled workers suitable primarily for manufacturing (we revert to the single final good as in the base case). For simplicity, we focus on the special case where there is complete segmentation between the two sectors; namely the manufacturing technology requires only unskilled workers and R&D only skilled workers.

Assume further that manufacturing of components requires one unit of unskilled labor per unit of output. In this event total production of components per unit time equals L. Components are priced according to (4) and final output according to (6), as in the base case, where w stands now for the wage rate of unskilled labor. This implies a constant wage rate for unskilled labor of $w = \alpha/L$. Denoting by H the supply of skilled labor, we choose labor units such that one unit of skilled labor develops one new product per unit time. Thus $\dot{n}_i = H$; in words, the flow of new components per unit time is constant and equals H. This implies that the number of components available at time t is $n_i(t) = (t - T_i)H$ for $t \in [T_i, T_{i+1})$ and that the length of the first phase is simply determined by $\Delta_1 = \eta\Delta$.

We restrict the discussion to parameter values that produce a two-phase cycle, such as in case 1 of section 3.3. In this event the no-arbitrage condition (10) implies that the evolution of the wage rate of skilled workers, denoted by w_H, satisfies

$$\dot{w}_H = \begin{cases} \rho w_H & \text{for } t \in [T_i, T_i + \eta\Delta), \\ \rho w_H - \dfrac{1-\alpha}{(t - T_i)H} & \text{for } t \in [T_i + \eta\Delta, T_{i+1}). \end{cases}$$

In addition, if it is profitable to develop components for the $(i-1)$th GPT at time T_i^-, then (see (21)):

$$w_H(T_{i+1}^-) = \frac{1-\alpha}{\rho H \Delta}(1 - e^{-\rho\eta\Delta}).$$

These provide us with a differential equation and an end condition. The solution to this system is

$$w_H(t) = \begin{cases} e^{\rho(t-T_i)} w_H(0) & \text{for } t \in [T_i, T_i + \eta\Delta), \\ e^{\rho(t-T_i)} \left\{ w_H(0) - \dfrac{1-\alpha}{H} [\Phi(\rho\eta\Delta) - \Phi(\rho(t - T_i))] \right\} \\ & \text{for } t \in [T_i + \eta\Delta, T_{i+1}), \end{cases} \quad (26)$$

where $w_H(0) = [(1-\alpha)/H][(1/\rho\Delta)e^{-\rho\Delta}(1 - e^{-\rho\eta\Delta}) - \Phi(\rho\Delta) + \Phi(\rho\eta\Delta)]$, and $\Phi(y) = \int_y^\infty (e^{-x}/x)\,dx$.

We can now describe the evolution of relative wages. Recalling that the wage rate of unskilled workers is constant here, the time trend of relative wages is fully determined by the wage rate of skilled workers, as described in (26). It follows that the relative wage rate of skilled workers rises during the first phase. However, during phase 2 the wage rate of skilled workers is always lower than its peak at the end of phase 1. Therefore the *relative* wage rate of skilled workers has to decline at least at the beginning of phase 2, and it may in fact decline all the way, until the arrival of a new GPT.[17]

Now let us turn to real wages. From (6) and the fact that the wage rate of unskilled workers is constant, it follows that the real wage rate of unskilled workers, w/p_Q, is constant in phase 1 and rising in phase 2.[18] On the other hand, the real wage rate of skilled workers, w_H/p_Q, rises during phase 1 because w_H rises and p_Q does not change. In phase 2 the real wage rate of skilled workers may decline.[19] We therefore see that in the early part of a long-run cycle, when productivity stagnates, so does the real wage of unskilled workers, while skilled workers gain higher real wages over time. Then, with the adoption of the newest general purpose technology, real income of unskilled workers turns around. In that phase both types of labor may experience rising real incomes, at least for some time. But further along the cycle skilled workers may suffer falling real wages.

As already mentioned, the changes in relative wages of skilled versus unskilled labor have received a great deal of attention in recent times, in view of the deterioration suffered by unskilled workers during the 1980s, and the large increase in their unemployment rate in countries with rigid labor markets. Two hypothesis compete for the explanation of these trends: (1) that the pressure on relative wages emanates from foreign competition, and especially from labor-intensive exports of less developed countries, and (2) that technological progress in the industrial countries has differentially affected the demand for labor with varying skills. Our analysis suggests that the observed changes in relative wages are con-

---

17. For example, when $\alpha = 0.5$, $\lambda = 1.5$, $\rho = 0.025$, $\Delta = 20$, $L = 1$, and $H = 0.3$, the wage rate of skilled workers declines at each point in time during phase 2.

18. An extension of the model can produce a declining real wage rate of unskilled workers during phase 1. For this we need to assume that both types of labor are used in manufacturing. Then, as innovators increase their demand for skilled labor with the appearance of a new GPT, skilled workers will be reallocated from manufacturing to R&D. As a result unskilled workers will collaborate with fewer skilled workers in manufacturing, and the marginal product of unskilled workers will decline, leading to lower wages of unskilled workers.

19. For the parameter values presented in footnote 15, the real wage rate of skilled workers rises during the early part of phase 2 and declines subsequently.

sistent with economies that are in the first phase of a long cycle, driven perhaps by a computer-based GPT.

Next consider real GDP, now given by

$$G = \frac{1}{p_Q}(1-\alpha+wL+w_H H) = L(1+w_H H)\lambda^k(t_k H)^{(1-\alpha)/\alpha} \quad \text{for } t \in [T_i, T_{i+1}),$$

where k is the index of the GPT actually used in production. We have $k = i-1$ and $t_k = \Delta$ for $t \in [T_i, T_i + \eta\Delta)$, and $k = i$ and $t_k = t - T_i$ for $t \in [T_i + \eta\Delta, T_{i+1})$. Notice that real GDP is homogeneous of degree $1/\alpha > 1$ in L and H (recall that w_H is inversely proportional to H, and hence $w_H H$ is invariant to factor endowments). Thus this economy exhibits aggregate economies of scale. As to growth, real GDP drops right at the beginning of a cycle (as in the base case), but then it grows over phase 1, contrary to the negative growth seen in the base case. The behavior of real GDP during the second phase is less clear: The simulations that we carried out for a particular parameter combination show that as in the base case, growth is significantly faster in the second phase (see figure

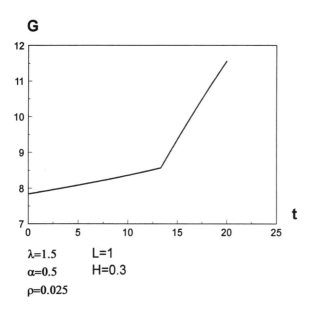

$\lambda=1.5 \quad L=1$
$\alpha=0.5 \quad H=0.3$
$\rho=0.025$

Figure 3.8
Output

3.8).[20] However, we have not been able to establish thus far how general this result is.

3.6 Concluding Remarks

The point of departure of this chapter is a view of the growth process in which the vague notions of technology-related "increasing returns" or "nonconvexities" that underlie a great deal of the new growth theory acquire a very concrete meaning: that of general purpose technologies fostering complementary advances in user sectors. The building blocks of our analysis can therefore be related to specific technological and historical processes (e.g., the electrification of manufacturing in the early twentieth century); similarly the outcomes of the analysis can be judged against that same historical and contextual backdrop.

The long-run equilibrium notion that emerges from our analysis is that of a recurrent cycle, associated with each new and ever-improving generation of GPTs. From a purely historical viewpoint, that seems to be a more compelling representation of actual processes than either stationary growth rates or convergence to saddle points. Within each cycle the analysis shows the centrality that complementary investments play in the growth process, and how the sequential and cumulative nature of such investments may induce different phases along each cycle, exhibiting very different features. Particularly striking is the initial phase of negative or slow growth, which results from the fact that there is a threshold level of complementary inputs that needs to be developed before the latest GPT can outcompete and displace the previous one.

The extensions sketched in section 3.6 suggest that this model may offer a suitable framework for the analysis of a series of important issues that arise in the interface of technology and economics. In particular, the model offers one possible way of addressing the troublesome decline in relative wages of unskilled workers during the 1980s. In fact the predicted dynamic behavior of wages in our extended model is consistent with the observed phenomenon, and it carries also an optimistic message: These trends are supposed to turn around once the latest GPT (computers?) reaches beyond its gestational stage and starts having a real impact on productivity. Clearly, though, the model developed here suffers from a

---

20. Notice that the pattern shown in figure 3.8 resembles the productivity path of electricity alluded to in section 3.4 (David 1991) more closely than the one in figure 3.5, which represents the base case.

series of readily apparent limitations, which we see as agenda for future research. Prominent among them is the fact that we have ignored the endogenous character of advances in the GPT itself and the associated (positive) feedback going from the user sectors to the GPT. This is an important part of the mechanism by which GPTs are thought to play their role of "prime movers," and its (hopefully transient) omission is due only to the modeling difficulties that we have encountered.

References

Bresnahan, T. F., and M. Trajtenberg. 1995. General purpose technologies: "Engines of growth"? *Journal of Econometrics* 65: 83–108, 1995.

Chandler, A. D. 1977. *The Visible Hand: The Managerial Revolution in America Business*. Cambridge: Harvard University Press.

Chou, C. F., and O. Shy. 1993. Technological revolutions and the gestation of new technologies. *International Economic Review* 34: 631–45.

David, P. A. 1991. Computer and dynamo: The modern productivity paradox in a not-too-distant mirror. In *Tehnology and Productivity: The Challenge for Economic Policy*. Paris: OECD, pp. 315–47.

Griliches, Z. 1957. Hybrid corn: An exploration in the economics of technological change. *Econometrica* 25: 501–22.

Grossman, G. M., and E. Helpman. 1991. *Innovation and Growth in the Global Economy*. Cambridge: MIT Press.

Helpman, E., and M. Trajtenberg. 1994. A time to sow and a time to reap: Growth based on general purpose technologies. Foerder Institute for Economic Research Working Paper 23–94.

Milgrom, P., and J. Roberts. 1990. The economics of modern manufacturing: Technology, strategy, and organization. *American Economic Review* 80: 511–28.

Romer, P. M. 1990. Endogenous technological change. *Journal of Political Economy* 98 (P. 2): S71–102.

4   Diffusion of General Purpose Technologies

Elhanan Helpman and
Manuel Trajtenberg

4.1 Introduction

Both historical evidence and theoretical analysis have brought forth the notion that general purpose technologies may play a key role in economic growth (chapter 3, this volume; see also David 1991; Bresnahan and Trajtenberg 1995). GPTs are characterized, first, by their pervasiveness in that they are used as inputs by a wide and ever expanding range of sectors in the economy. As a GPT diffuses, it fosters complementary investments and technical change in the user sectors, thereby bringing about sustained and pervasive productivity gains. The steam engine during the first industrial revolution, electricity in the early part of this century, and microelectronics in the past two decades are widely thought to have played such a role.

In chapter 3 we incorporated the notion of GPTs into a growth model and explored the economywide dynamics that a GPT may generate. We assumed that advances in the GPT are exogenous and that they occur at fixed intervals over time. We found that each time a new GPT appears, it generates a cycle consisting of two distinct phases. The first phase is the "time to sow," when resources are diverted to the development of complementary inputs that take advantage of the new GPT. During this initial stage output and productivity experience negative growth, and the real wage stagnates. The "time to reap" comes in the second phase, after enough complementary inputs have been developed and it becomes worthwhile to switch to manufacturing with the new, more productive GPT. As a consequence there is a spell of growth, with rising output, real wages, and profits. As new GPTs appear, these cycles are repeated.

Helpman: Harvard University, Tel Aviv University, and CIAR. Trajtenberg: Tel Aviv University and CIAR. We are grateful to the National Science Foundation for financial support and to two referees for comments.

In chapter 3 we treated the order of adoption in a crude way. However, the order of adoption, the "lateral" linkages between adopting sectors, and the resulting speed and extent of the diffusion process, are all crucial aspects of the dynamic process associated with a GPT and hence important determinants of the rate of growth induced by GPTs.

Economists have long been aware of the fact that technical advances are surprisingly slow in spreading and hence that the time profile of the diffusion process is key in determining the realization of the benefits, social and private, from technical change (Griliches 1957; Mansfield 1968). However, diffusion studies have typically looked at individual innovations, whereas in this chapter we seek to analyze the *economywide* implications of the diffusion of "macro" innovations in the form of GPTs.

The basic framework is that of chapter 3, except that here we assume that there are a multiplicity of sectors (or final goods), each of them characterized by a set of parameters that account for the fact that some adopt earlier than others, thus generating a diffusion process. The parameter set includes (1) a technology parameter representing the productivity advantage of the new GPT (relative to the old one) for each sector, (2) a "historical stock" parameter indicating the number of components developed for the old GPT, (3) a demand parameter (the spending share of the sector), and (4) an R&D parameter (how expensive it is to develop new, complementary components for the specific sector). The first and second parameters jointly determine the number of new components that have to be developed in order to make it worthwhile to switch to the new GPT. In other words, they determine the height of the "switching threshold."

Assuming an exogenous ordering of sectors, we trace the dynamic trajectory of the economy as one sector after the other adopts the new GPT. Each generates a cycle similar in nature to the economywide cycle found in chapter 3: Each cycle is made of two phases, the first depressing real income; the second pushing it up as the sector adopts the new GPT in manufacturing. Hence it realizes its benefits in the form of enhanced productivity. As all the sectors adopt and the economy moves toward its steady state, a "second wave" occurs: All sectors engage anew in R&D, whose benefits occur immediately (since all sectors are already producing with the new GPT), and bring about a spell of sustained and widespread growth.

The picture that emerges is as follows: The initial stages in the spread of a new GPT render a process of intermittent growth as early adopters make the required complementary investments and laggards keep producing with the old GPT. It is only when a critical mass of sectors has adopted, and real income risen enough, that pioneers and laggards alike are induced to make further complementary investments that pay off right away, hence generating a spell of sustained growth. We also analyze the welfare implications of the order of adoption (by way of numerical simulations) and find that as intuition would suggest, welfare is higher if the sectors being first to adopt have low switching thresholds, low R&D costs, or high spending shares (the demand parameter).

The setting just described may render a large number of alternative equilibria, since many different sequences of sectors (in terms of order of adoption) may be plausible. Of course, if one were to allow for expectations other than perfect foresight (as assumed here), many more equilibrium trajectories would become feasible. This is by no means a deficiency of the model but a reflection of the complexity and inherent indeterminacy of the issue at hand. This is clearly manifested in the last section of the paper, where we sketch the early diffusion of the transistor, the first embodiment of the leading GPT of our times, semiconductors.

In particular, we seek to characterize both the early adopters of the transistor (hearing aids and computers) and the laggards (telecommunications and automobiles) in the parameters of the model. Not surprisingly, virtually all parameters mentioned above were advantageous for early adopters. Of particular importance seems to have been the plain inadequacy of the incumbent technology for hearing aids and computers vis à vis the new if still rough GPT. As to the laggards, the decisive impediment to early adoption was an extremely large historical stock, in the form of the sheer size of the system and the complex interdependence of components for telecommunications. For automobiles it referred to the huge stock of manufacturing plants and the disruption that the introduction of the new GPT could bring upon it. Interestingly, the errors of forecast made initially with respect to the laggards were due to the fact that these predictions were based primarily on the expected relative technological advantage of the new GPT, and not on the basis of the height of the threshold that these sectors would have to overcome in order to bring themselves to adopt.

4.2 Building Blocks

There are m sectors. Final output of sector i, Q_i, is produced with an assembly of components $x_i(h)$, $h \in [0, n_i]$, where n_i stands for the number of different components available to the sector. These components have to fulfill two conditions: They have to be compatible with the particular GPT in use by the sector (i.e., components developed for previous GPTs would not work with a subsequent GPT), and they are specific to the sector itself.

4.2.1 Production

The economy is in a steady state, manufacturing with an old GPT. We begin the analysis at time $t = 0$, when a new more productive GPT arrives unexpectedly. The production function of sector i with the new GPT is

$$Q_i = G_i D_i, \qquad G_i > 1, \tag{1}$$

$$D_i = \left[\int_0^{n_i} x_i(h)^\alpha dh \right]^{1/\alpha}, \qquad 0 < \alpha < 1.$$

The elasticity of substitution between any two components is thus $1/(1-\alpha)$, the same in all sectors. But at time zero no components are available for the new GPT. Therefore $n_i(0) = 0$ for all sectors, and no production can take place with the new technology.

We can allow a variety of specifications for the old technologies without substantially affecting the analysis. It proves, however, convenient to think about the old GPT as having production functions such as (1) with $G_i = 1$ for every sector. In this event the advantage of the new GPT over the old one is represented by the fact that $G_i > 1$ for the new technology. The larger G_i the bigger the advantage of the new technology for sector i. Next let the number of components available in each sector for the old technology be given. Then manufacturers of final outputs, who are competitive in the product and input markets, can calculate marginal costs for the old technology. They continue to use the old technology as long as these marginal costs are lower than the marginal costs that result from the use of the new technology. But since at time zero there are no components to go with the new technology, marginal costs are necessarily lower for the old technology. Therefore the old technology remains in use at least for some time.

4.2.2 Pricing

To supply a component for the new technology in sector i, a firm has to invest resources in R&D in order to develop its own unique brand. This invention is protected by a patent, and each such firm engages in monopolistic competition with the other suppliers of components to that sector. All such firms and corresponding components are symmetric, all components can be produced with one unit of labor per unit output, and hence marginal costs equal the wage rate, w, which we normalize as $w = \alpha$. Thus profit-maximizing producers of components, who face demand functions with a constant price elasticity $1/(1 - \alpha)$, price them equally;

$$p_i(h) = \frac{w}{\alpha} = 1 \quad \text{for all } h \in [0, n_i], \quad \text{all } i = 1, \ldots, m. \tag{2}$$

The last equality results from our choice of units (i.e., the wage rate equals α). Under these circumstances all components for the ith sector are employed in equal quantities x_i. As a result profits per new component in sector i are

$$\pi_i = \frac{(1 - \alpha)E_i}{n_i}, \tag{3}$$

where E_i is spending on good i. This formula applies when the new GPT is used in the sector; otherwise, profits equal zero. From (1) and the fact that all components for a given sector are used in equal quantities, it also follows that

$$Q_i = G_i n_i^{(1-\alpha)/\alpha} X_i, \tag{4}$$

where $X_i = n_i x_i$ is the total employment of inputs. The same type of relationships apply to the old technology.

4.2.3 Resource Constraint

The total amount of labor available in the economy is L, which has to be allocated between the production of components and R&D, where R&D is designed to develop components for the new GPT. Therefore the resource constraint is

$$\sum_{i=1}^{m} X_i^o + \sum_{i=1}^{m} X_i + \sum_{i=1}^{m} a_i \dot{n}_i = L,$$

where X_i^o is the output of old type components for sector i, X_i is the output of new type components for sector i, a_i is the labor requirement per unit invention of new components for sector i, and $\dot{n}_i$ is the flow of new components invented for sector i per unit time. We assume for the time being that the R&D coefficients a_i are constant. Since the production of final outputs is competitive and there are constant returns to scale for given components, it follows that $X_i^o + X_i = E_i$ and therefore that $\sum_{i=1}^m X_i^o + \sum_{i=1}^m X_i = E$, where $E = \sum_{i=1}^m E_i$ represents aggregate spending. In this event the resource constraint can be represented as

$$E + \sum_{i=1}^m a_i \dot{n}_i = L. \tag{5}$$

In addition aggregate profits of component manufacturers (new and old) equal $(1-\alpha)E$. Therefore national income is given by

$$Y = \alpha L + (1-\alpha)E, \tag{6}$$

measured in the same units as wages. Divided by α, it gives real income in labor units.

In order to take advantage of the new technology, each sector has to develop new components, and enough of them so as to make it worthwhile to switch to the new GPT. We denote by n_{ic} the number of components that have to be developed for sector i in order to induce manufacturers of Q_i to switch to the new technology.[1]

4.2.4 Profits and the Market Value of Firms

At each point in time consumers maximize a Cobb-Douglas utility function,

$$u = \prod_{i=1}^m Q_i^{\beta_i}, \quad \sum \beta_i = 1, \tag{7}$$

implying that $P_i Q_i = E_i = \beta_i E$, where P_i is the price of Q_i. Combined with (3), this relationship implies that

$$\pi_i = \frac{(1-\alpha)\beta_i E}{n_i}. \tag{8}$$

The stock market value of each such firm at time t is

$$v_i(t) = \int_t^\infty e^{-[R(\tau)-R(t)]} \pi_i(\tau) \, d\tau, \tag{9}$$

---

1. Unit costs in the production of Q_i are $c_i = X_i/Q_i = 1/G_i n_i^{(1-\alpha)/\alpha}$ with the new technology and $c_i^o = X_i^o/Q_i^o = 1/(n_i^o)^{(1-\alpha)/\alpha}$ with the old (see (4)). Therefore the new one becomes profitable whenever $n_i \geq n_{ic} \equiv n_i^o/(G_i)^{\alpha/(1-\alpha)}$.

where $R(\tau)$ stands for the discount factor from time τ to zero. We note that this valuation equation implies the no-arbitrage condition

$$\frac{\pi_i}{v_i} + \frac{\dot{v}_i}{v_i} = r, \tag{10}$$

where $r = \dot{R}$ is the instantaneous interest rate.

It pays to develop components for sector i at time t only if the value of a firm is at least as high as the cost of R&D, αa_i. We assume free entry into the R&D business. Therefore, whenever new components are developed for sector i, the free entry condition implies that

$$v_i = \alpha a_i, \tag{11}$$

whereas no development takes place when $v_i < \alpha a_i$.

4.2.5 Intertemporal allocation of spending

Consumer preferences at time zero are

$$U(0) = \int_0^\infty e^{-\rho \tau} \log u(\tau) \, d\tau, \tag{12}$$

where u is given in (7) and ρ is the subjective discount rate. Each consumer maximizes this preference function subject to a budget constraint in which the present value of spending equals the present value of labor income plus the value of firms. The first order conditions of this optimization problem imply that

$$e^{-\rho t} E(0) = e^{-R(t)} E(t). \tag{13}$$

Differentiating this condition with respect to time results in the familiar differential equation for aggregate spending

$$\frac{\dot{E}}{E} = r - \rho.$$

Namely spending grows at a rate that equals the difference between the interest rate and the subjective discount rate. Combined with the no-arbitrage condition (10), this implies that

$$\frac{\dot{E}}{E} = \frac{\pi_i}{v_i} + \frac{\dot{v}_i}{v_i} - \rho \quad \text{for every } i. \tag{14}$$

An important property that emerges from (13) is that aggregate spending is a continuous function of time if and only if the discount factor never jumps. We will see that the discount factor indeed cannot jump. Therefore spending varies continuously on the equilibrium trajectory, except of course at time zero when the new GPT arrives.

4.3 Basic Dynamics

A key question in the diffusion of a new general purpose technology concerns the sequence in which the technology is adopted by various sectors. For the moment, however, suppose that this sequence is given. In this event it is convenient to number the sectors according to the order of adoption; that is, sector $i = 1$ is the first to adopt the new GPT, sector $i = 2$ is the second to adopt it, and so on, until $i = m$. We will discuss the economic factors that determine the order of adoption in section 4.4.

4.3.1 *Spending*

Sector i replaces the old technology with the new one only if the number of components available for the new GPT exceeds n_{ic}. This means that there has to be a time interval in which new components are developed for sector i but in which manufacturers of final output still use the old technology. In this time interval manufacturers of the new components have no buyers and therefore do not produce them, which implies that they have zero profits.[2] In addition, as long as new components are being developed, the free entry condition (11) holds, implying that the value of new component firms is constant in this industry. But zero profits and a constant value of firms imply that (see (14))

$$\frac{\dot{E}}{E} = -\rho. \qquad (15)$$

Namely, as long as there is an industry that develops new components but in which the old GPT remains in use, aggregate spending declines at a constant rate that equals the subjective discount rate. It results from the fact that during such time periods the interest rate equals zero (see the no-arbitrage condition (10)).

---

2. Note that this result stems from the fact that there is monopolistic competition in the components sector. As a result component producing firms cannot coordinate their prices in order to induce component buyers to switch to the new technology.

The intuition behind this result is that developers of components who expect to make zero profits for some time prefer to postpone R&D investment as long as the interest rate is positive. Therefore the interest rate has to be zero in order to induce them to innovate.

On the other hand, if sector i has already switched to the new technology (i.e., $n_i \geq n_{ic}$), then profits per component are given by (8). In addition, whenever new components for the sector are being developed the free entry condition implies that the value of a component firm is constant. As a result (8), (11), and (14) imply that

$$\frac{\dot{E}}{E} = \frac{(1-\alpha)\beta_i E}{\alpha a_i n_i} - \rho. \tag{16}$$

If more then one sector that has switched to the new technology develops components at the same time, then this equation holds simultaneously for each one of them. Clearly this can happen only if the ratio $\beta_i/a_i n_i$ is the same in all of them. It is also clear from a comparison of (15) with (16) that it is impossible to have simultaneous development of components in a sector that has not yet switched to the new technology and a sector that has.

4.3.2 Research and Development

There are three possibilities at a point in time: Firms do not invest in R&D, new components are being developed for one sector only, new components are being developed for more than one sector. In the first case the resource constraint (5) implies constant spending; namely $E = L$. In the second case it implies that

$$a_i \dot{n}_i = L - E, \tag{17}$$

where R&D is taking place only in sector i. This is a differential equation for the number of brands developed in that sector. In the third case the resource constraint implies that

$$\sum_{i \in I} a_i \dot{n}_i = L - E, \tag{18}$$

where I is the set of sectors in which R&D takes place. But we have seen that whenever more than one sector innovates, then either all the innovating sectors have not yet switched to the new technology or all of them have (but it is not possible for only some of them to have switched).

In the latter case, which is the only case of interest, we must have (see (16))

$$\frac{a_i n_i}{\beta_i} = \gamma_I \quad \text{for all } i \in I. \tag{19}$$

Clearly the case in which only one sector innovates is a special case in which the set I is a singleton; namely $I = \{i\}$.

Why is the case in which innovation takes place in a number of sectors that have not yet switched to the new technology of no particular interest? Suppose that this is indeed the case. Then, if one of these sectors reaches the critical number of components that induces a switch to the new technology and this sector continues to developed components after that point in time, then the interest rate has to be positive after the switch.[3] As a result all other sectors that developed with it, but who have not yet reached the critical number of components that induces a technology switch, regret the early investment in R&D (they would be better off postponing the investment in product development). But this type of regret cannot happen on an equilibrium trajectory. Therefore either all of them reach the critical number of components simultaneously or the sector that reaches this point first stops further product development. Both are knife-edge cases that we assume away.

4.3.3 *Continuous Discounting*

Consider a point in time t at which some R&D takes place. It is apparent from the valuation of firms in (9) that v_i is a continuous function of time. Therefore postponing at t the development of a product by a negligible amount of time has a negligible effect on the reward of doing R&D. At the same time the postponement of product development by a negligible amount of time has a negligible effect on R&D costs if and only if the discount factor $R(\cdot)$ is continuous to the right of t. For example, if the discount factor is discontinuous to the right of t and $R(t^+) > R(t)$, it is

---

3. Suppose that the interest rate is zero after the switch. Then the no-arbitrage condition (10) implies that the value of a components firm that makes positive profits has to decline. On the other hand, as long as there is product development, the value of a firm has to be constant and equal to the costs of R&D (see (11)), a contradiction. Therefore the interest rate cannot be zero. The economic reason for the nonzero interest rate is that if the interest rate were zero and the value of a profit-making firm were expected to remain constant at the level of its R&D costs, then investment in such firms would be infinitely profitable.

preferable to invest in product development a moment after t, and if $R(t^+) < R(t)$, then it is strictly preferable to invest in R&D at t as compared with a moment later. A similar argument can be made with respect to discontinuities of $R(\cdot)$ to the left of t. It follows that there can be no investment in product development over an entire interval of time unless the discount factor is continuous on this time interval. We seek equilibrium trajectories on which there is research and development from the moment of arrival of the new GPT and until the economy reaches a new steady state.[4] On all such trajectories the discount factor has to be a continuous function of time. It then follows from the consumers' first-order condition (13) that on all such trajectories aggregate spending is a continuous function of time.

4.3.4 Sectoral Wave

As indicated at the beginning of this section, for the time being we take as given the order of adoption of the new technology. In this event we may think of a typical time interval over which firms develop components for sector i, which follows a time interval over which R&D was directed toward product development in the preceding sectors. Suppose that the development of components for sector i begins at time T_i and continues until T_{i+1}, with no components for other sectors being developed in the meantime. Also suppose that in the time interval $[T_i, T_{i+1})$ manufacturers of final output Q_i have switched at some point to the new technology. Then we can divide this time interval into two phases: In phase 1 components for i are being developed while the old technology is in use, and in phase 2 components for i are being developed while the new technology is in use. The switch from phase 1 to 2 occurs when n_{ic} components become available.

In phase 1 aggregate spending evolves according to (15) while the number of components evolves according to (17). The phase begins with $n_i = 0$ and ends with $n_i = n_{ic}$. It is characterized by declining spending and a rising number of components, as depicted in figure 4.1. The initial level of spending $E(T_i)$ equals the level of spending at time T_i^-, because spending is continuous in time. Given this initial level of spending our differential equations determine the level of spending over the entire phase, and

---

4. It is very unlikely that periods in which there is no product development will follow and precede periods with product development. This point will become apparent when we discuss the determination of the initial value of spending.

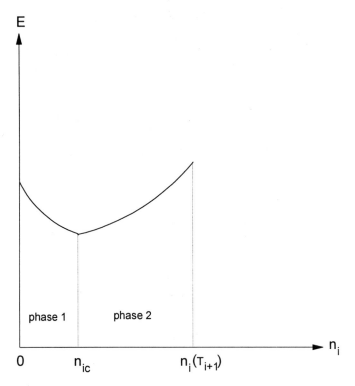

Figure 4.1
Sectoral wave

in particular $E(T_{ic})$, where T_{ic} is the point in time at which phase 1 ends. The phase ends when the number of available components equals n_{ic} and manufactures of final output switch to the new technology.

By continuity, phase 2 begins with the spending level $E(T_{ic})$ and the number of components n_{ic}. Starting with these initial conditions spending in phase 2 evolves according to (16), while (17) continues to be the differential equation for the number of components. The number of components continues to rise, but spending may rise or decline. Figure 4.1 depicts a phase-2 trajectory on which spending rises. Importantly, the end of this phase determines a number of components $n_i(T_{i+1})$ and a level of spending $E(T_{i+1})$. This spending level serves as the initial spending level for the dynamics that follow.

The two phases combined generate a sectoral wave of the type depicted in figure 4.1. It remains to be seen how such sectoral waves fit in into equilibrium trajectories.

4.4 Equilibrium Trajectories

Building blocks for an analysis of economic dynamics have been developed in the previous section. We now use them to construct equilibrium trajectories. For this purpose first note that in economies of this type perfect foresight leads to a steady state in the long run. We therefore seek trajectories that consist of two-phase sectoral waves that converge to a steady state. To this end we first describe the steady state.

4.4.1 Steady State

In a long-run steady state there is no product development. Therefore by the resource constraint (5) aggregate spending equals the labor force:[5]

$$\tilde{E} = L. \tag{20}$$

Under these circumstances (13) implies that the interest rate equals the subjective discount rate. We consider economies in which all m sectors adopt the new technology. Therefore (8), (9), and (20) together with the fact that the interest rate equals ρ imply

$$\tilde{v}_i = \frac{(1-\alpha)\beta_i L}{\rho \tilde{n}_i} \quad \text{for all } i = 1, 2, \ldots, m. \tag{21}$$

Clearly this value of a firm cannot exceed αa_i; if it were, it would be profitable to further invest in product development. It follows that $\tilde{n}_i \geq (1-\alpha)\beta_i L/\rho \alpha a_i$. On the other hand, if $\tilde{n}_i > (1-\alpha)\beta_i L/\rho \alpha a_i$, then it cannot be that this number of components has been attained by investment in R&D in the last phase of the equilibrium trajectory, just prior to its reaching the steady state. To see why, observe that as long as there is R&D the resource constraint implies that aggregate spending is smaller than L. Therefore aggregate spending is rising during this last phase in order to reach $\tilde{E} = L$, while the differential equations that drive the system consist of (16) for the spending level (for $i \in I$) and (18) for the number of components, with the side condition (19). Since E is rising, it follows from (16) that $n_i < (1-\alpha)\beta_i L/\rho \alpha a_i$ during this phase, which proves our claim. But, if components for sector i are not developed during this last phase, then their development ceased at some earlier point in time, say at T_{i+1}. If so, then (8), (9), and (13) imply that at that point in time the value of a components firm in industry i was $(1-\alpha)\beta_i E(T_{i+1})/\rho \tilde{n}_i$ and that this value equaled product development costs αa_i, implying that $\tilde{n}_i =$

---

5. We use tildes to denote steady-state values.

$(1-\alpha)\beta_i E(T_{i+1})/\rho\alpha a_i$. Because there was product development at time T_{i+1}, however, aggregated spending was smaller than L (due to the resource constraint), implying that $\tilde{n}_i < (1-\alpha)\beta_i L/\rho\alpha a_i$, a contradiction. We therefore conclude that

$$\tilde{n}_i = \frac{(1-\alpha)\beta_i L}{\rho\alpha a_i} \quad \text{for all } i = 1, 2, \ldots, m, \tag{22}$$

and that there is product development in all sectors when the economy converges to the steady state.

4.4.2 Second Wave

Our analysis identified a number of features in and around the steady state. Importantly every sector has to engage in product development when the economy converges to the steady state. This means that all sectors that stopped developing new components at some earlier point in time are bound to renew product development during this last round. As a result *there has to be a second wave of product development for each sector, except for the last one.*

The fact that each sector has to invest in R&D when the economy approaches the steady state proves that there has to be a second wave of product development, but it does not exclude the possibility that some sectors will experience more than two. The dynamics of, second, third, and subsequent, waves of product development are described by (16) for spending and by (18) for the number of components, with the side condition (19), where I is the set of sectors that jointly engage in R&D. The same system can be represented in an alternative form that suppresses the side condition:

$$\frac{\dot{E}}{E} = \frac{(1-\alpha)E}{\alpha\gamma_I} - \rho, \tag{23}$$

$$\dot{\gamma}_I \sum_{i \in I} \beta_i = L - E. \tag{24}$$

During the convergence to a steady state, this system applies with $I = \{1, 2, \ldots, m\}$, and therefore $\sum_{i \in I} \beta_i = 1$.

Figure 4.2 describes the unique saddle path of (23) and (24) that leads to a steady state. On this trajectory spending rises and so does the number of components in each sector. The equilibrium trajectory hits this saddle path at some point in time and coincides with it from this point on. In order for all sectors to engage in R&D simultaneously, however, they

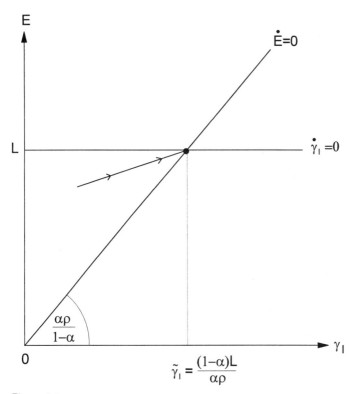

Figure 4.2
Steady state

have to begin R&D activities in a suitable order, following the preceding sectoral waves. This order is endogenous, given the order in which the sectoral waves evolve.

To determine the order in which sectors join the last round of product development, suppose that sector i ended the most recent period of product development at time $\hat{T}_i$ with $\hat{n}_i = n_i(\hat{T}_i)$ components. As a result sector i does not engage in product development from $\hat{T}_i^+$ until it joins the last round of development. Given the values of $\hat{n}_i$ for all $i = 1, 2, \ldots, m-1$, we can identify the order and timing by means of the following:

Last Round Procedure

1. Start from the steady state.
2. Run backward the differential equations (23) and (24) for $I = \{1, 2, \ldots, m\}$, and calculate $E(t)$ and $n_i(t)$. The number of components declines for each sector in this backward calculation.

3. Let i_{m-1} be the first sector for which $n_i(t) = \hat{n}_i$, and let this happen at time $\bar{T}_{i_{m-1}}$. Then this is the last sector to join the last round of development.

4. Starting with the spending level and the number of products that were calculated in the previous step for $t = \bar{T}_{i_{m-1}}$, repeat step 2 with $I = \{1, 2, \ldots, m-1\} - \{i_{m-1}\}$ (i.e., by dropping sector i_{m-1} from the set of sectors that are active in R&D). Stop this calculation at time $\bar{T}_{i_{m-2}}$ when a second sector, i_{m-2}, satisfies $n_i(t) = \hat{n}_i$. Sector i_{m-2} is the second to the last to join the last round of development.

5. Proceed in similar fashion to identify the order of all remaining sectors.

At the end of the process we obtain a perturbation $(i_1, i_2, \ldots, i_{m-1})$ of $(1, 2, \ldots, m-1)$, which represents the order in which the sectors begin the last round of product development; sector i_1 is the first to begin this last round while i_{m-1} is the last. We denote by $\iota(i)$ the order of sector i in this perturbation. Namely $\iota(i) = j$ if and only if $i_j = i$. This order depends, however, on the values of $\hat{n}_i$, $i = 1, 2, \ldots, m-1$, which are also endogenous. It is therefore necessary to examine the entire trajectory in order to identify the equilibrium values of $\hat{n}_i$.

4.4.3 Entire Trajectory

Our first task in describing the equilibrium trajectory is to characterize the equilibrium values of $\hat{n}_i$. For simplicity we restrict this discussion to trajectories on which each industry experiences a sectoral wave and renews R&D only during the last round of product development. Namely we abstract from situations in which an industry experiences a sectoral wave, develops new products at some later stage (or stages) as well, and finally joins the last round of product development. Such equilibrium trajectories cannot be excluded, however, and we will discuss them briefly later on.

So suppose that we run (23) and (24) backward, using the five steps of the Last Round Procedure. At the end of the process we obtain an order in which sectors join the last round of product development, i_j, $j = 1, 2, \ldots, m-1$, and the points in time at which they join in, $\bar{T}_{i_j}$. Now proceed to the next set of steps:

Fixed Point Procedure

1. Start at $\bar{T}_1$ with $E(\bar{T}_1)$ and $n_m(\bar{T}_1)$. At this point in time the last sector, m, completes its sectoral wave.

2. Run backward the differential equations (16) and (17) for $i = m$ until $n_m(t) = n_{mc}$. This identifies phase 2 of the sectoral wave in m. In particular, it identifies the point in time at which phase 2 begins, T_{mc}, and the spending level $E(T_{mc})$. Naturally $n_m(T_{mc}) = n_{mc}$.

3. Starting with $E(T_{mc})$ and $n_m(T_{mc}) = n_{mc}$, run backward the differential equations (15) and (17) for $i = m$ until $n_m(t) = 0$. This identifies phase 1 of the sectoral wave in m. In particular, it identifies the point in time at which phase 1 begins, T_m, and the spending level $E(T_m)$. This point in time also represents the end of phase 2 in sector $m - 1$.

4. Sector $m - 1$ ends its sectoral wave with $n_{m-1}(T_m) = \hat{n}_{m-1}$ components, and it joins the last phase of product development at time $\bar{T}_{\iota(m-1)}$. Between time T_m and $\bar{T}_{\iota(m-1)}$ it does not invest in R&D. It therefore follows from the valuation of firms (9) and the free entry condition (11) that $\hat{n}_{m-1}$ has to satisfy[6]

$$\hat{n}_{m-1} = \frac{\beta(1-\alpha)E(T_m)}{\rho \alpha a_{m-1}} \frac{1 - e^{-[\bar{T}_{\iota(m-1)} - T_m]}}{1 - [E(T_m)/E(\bar{T}_{\iota(m-1)})]e^{-[\bar{T}_{\iota(m-1)} - T_m]}} . \tag{25}$$

If this condition does not hold, it means that the Last Round Procedure did not use the equilibrium value of $\hat{n}_{m-1}$.

5. Starting with $E(T_m)$ and $n_{m-1}(T_m) = \hat{n}_{m-1}$, repeat steps 2–4 (with suitably modified indexes) for sector $i = m - 1$, then for $i = m - 2$, and so on, until $i = 1$. This traces out the remaining parts of the equilibrium trajectory.

It is clear from these procedures that they can be used to find the equilibrium values of $\hat{n}_i$ as a fixed point of a mapping. Given a vector $(\hat{n}_1, \hat{n}_2, \ldots, \hat{n}_{m-1})$ our Last Round and Fixed Point procedures map this vector into another vector by means of (25). A fixed point of this mapping identifies the equilibrium values.

Figure 4.3 depicts an equilibrium trajectory for a three-sector economy, combining the features of the sectoral wave in figure 4.1 with the convergence to a steady state in figure 4.2. The aggregate number of components is measured on the horizontal axis. Product development begins in sector 1. Its first phase lasts until n_{1c} components have been developed,

---

6. The valuation equation (9) implies that for every t_1 and $t_2 > t_1$,

$$v_i(t_1) = \int_{t_1}^{t_2} e^{-[R(t)-R(t_1)]} \pi_i(t) \, dt + e^{-[R(t_2)-R(t_1)]} v_i(t_2).$$

In our case $t_1 = T_m$ and $t_2 = \bar{T}_{\iota(m-1)}$, the free entry condition (11) holds at the two end points, and profits satisfy (8) at each point in time. Combined with the consumer's first-order condition, (13), these conditions and the valuation equation imply (25).

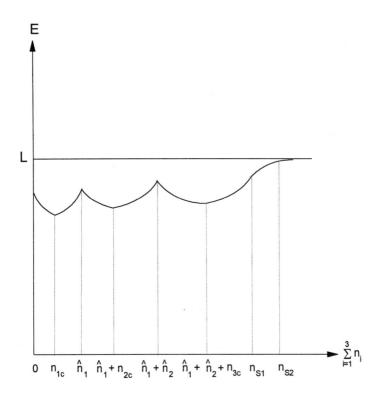

n_{S1} = one sector joins the second wave

n_{S2} = remaining sector joins the second wave

Figure 4.3
Equilibrium trajectory

whereupon its second phase begins, after manufacturers of final output switch to the new technology. The second phase ends when there are $\hat{n}_1$ components in sector 1. Then sector 2 begins the first phase of its sectoral wave. This one lasts until there are n_{2c} components to go with the new technology in sector 2. And so on. The third sector completes its sectoral wave when there are $n_{S1} = \sum_{i=1}^{3} \hat{n}_i$ components. At this point sector i_1 (which is either sector 1 or 2) joins the last round of product development. Sectors 3 and i_1 continue to invest in R&D until the number of components equals n_{S2}, at which point the remaining sector i_2 also joins the last round of development. From this point on all three sectors engage in R&D until the economy approaches the steady state, in which product development ceases.

Figure 4.3 describes an equilibrium trajectory on which the second wave is also the last round of product development, which is the case on which our analytical part has focused. To ensure that this is indeed an equilibrium, it also has to satisfy $v_i(t) \leq \alpha a_i$ at each point in time for every sector. Naturally this holds as an equality during periods of product development for sector i. If we calculate a candidate equilibrium by means of our procedures and it turns out that this inequality does not hold for some sector j in some time interval, it means that the second wave is not the last one for this sector. It is then necessary to search for equilibrium trajectories in which sector j starts a second wave of product development before the last, sometime during phase 2 of a sectoral wave in some other sector with a higher index. As a consequence sector j experiences more than two waves of product development. We will not elaborate a procedure for calculating this type of trajectories.

4.4.4 Real Wages and Real Income

By our normalization the wage rate equals α at each point in time, while income, Y, equals $\alpha L + (1 - \alpha)E$ (see (6)). In order to obtain the real wage rate and real income, we need to deflate these quantities by a suitable price index. From individual preferences (7), the price index of real consumption can be represented by

$$P = P_0 \prod_{i=1}^{m} P_i^{\beta_i},$$

where P_0 is some constant and P_i is the price of final output in sector i. On the other hand, it follows from (4) and the definition of n_{ic} (see footnote 1) that the price of final output in i, which equals unit costs, is given by

$$P_i = \begin{cases} G_i^{-1} n_{ic}^{-(1-\alpha)/\alpha} & \text{for } n_i \leq n_{ic}, \\ G_i^{-1} n_i^{-(1-\alpha)/\alpha} & \text{for } n_i \geq n_{ic}. \end{cases}$$

By a suitable choice of P_0 we can therefore represent the price index of consumption as

$$P = \prod_{i=1}^{m} \phi_i(n_i)^{\beta_i}, \tag{26}$$

$$\phi_i(n_i) = \begin{cases} n_{ic}^{-(1-\alpha)/\alpha} & \text{for } n_i \leq n_{ic}, \\ n_i^{-(1-\alpha)/\alpha} & \text{for } n_i \geq n_{ic}. \end{cases}$$

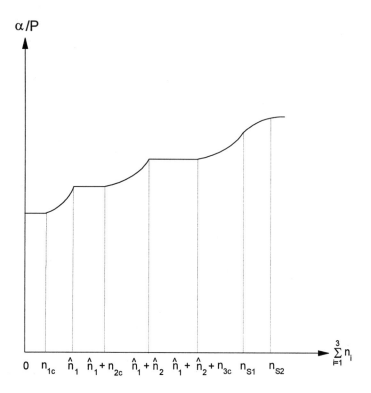

n_{S1} = one sector joins the second wave

n_{S2} = remaining sector joins the second wave

Figure 4.4
Real wages

The real wage rate α/P on the trajectory from figure 4.3 is depicted in figure 4.4. It remains constant at its initial level during the first phase of the first sectoral wave, and it rises steadily during phase 2 of this sectoral wave. This stems from the fact that as long as sector 1 has not switched to the new technology, the purchasing power of wages remains constant. But once it has switched, the price of good 1 declines as long as additional components are developed for sector 1. As a result real wages rise. A similar pattern of real wages evolves in each subsequent sectoral wave. In the final round, when all sectors have switched to the new technology, real wages rise steadily as long as R&D is positive. Real wages stabilize when the economy settles down on the steady state. Our model therefore

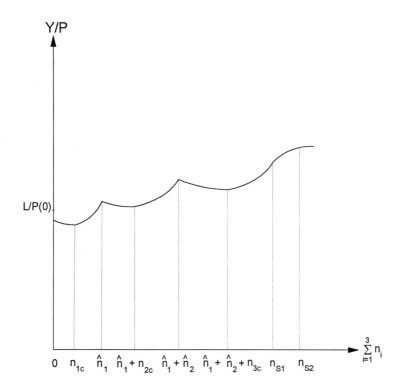

n_{S1} = one sector joins the second wave

n_{S2} = remaining sector joins the second wave

Figure 4.5
Real income

predicts periods of rising real wages, followed by periods of stagnation, as the new technology diffuses across sectors.

Figure 4.5 depicts the pattern of real income, $Y/P = [\alpha L + (1-\alpha)E]/P$, measured in real consumption units. The arrival of the new GPT leads to a drop in real income, as resources shift from manufacturing to R&D. This is similar to the result in chapter 3. From this point on real income declines continuously during phase 1 of the first sectoral wave and rises during phase 2. The same pattern is exhibited in subsequent sectoral waves until the last round in which real income rises and the economy arrives at the steady state. During the first phase of each sectoral wave prices of final goods are constant, but demand for them declines. This falling demand, which is accompanied by rising investment in R&D, reduces real income.

The reason is that due to the monopolistic price distortion in the production of intermediates, the employment of resources in manufacturing is too small to begin with. In phase 2 of a sectoral wave aggregate demand for final goods rises, and one of the prices declines. On account of both these trends real income rises. Since in the steady state real income is higher than in the initial steady state, it follows that the economy experiences average productivity growth, but this positive average growth is driven by cycles of declining and rising productivity levels as the use of the new technology spreads across sectors.

4.5 Order of Adoption

Our analysis of equilibrium trajectories assumed a known order in which the new technology spreads across sectors. Given this order, there are sectoral waves and a final round of across-the-board product development.

In this section we raise two questions. First, what features determine the order of adoption? And second, if there are multiple equilibria that differ in this order, how are they ranked according to relative efficiency?

4.5.1 *Multiple Equilibria*

The first thing to note is that there are multiple equilibrium trajectories in economies of this sort. To see why, observe that the profitability of developing a component for a given sector depends on expectations as to whether others will also develop new components for this industry. If one potential component developer expects that others will not invest, he understands that the sector will not adopt the new technology even if he does develop his particular brand. As a result he will not be able to sell any output and will not be able to cover R&D costs. Hence he does not develop the component.

This argument makes clear a critical point; R&D requires coordination of expectations among potential developers. If all believe that there will be sufficient development to induce the manufacturers of final output to switch to the new technology, then they invest in R&D and thereby justify these expectations. If, however, they do not believe that there will be enough product development, then none of them invests in R&D, and these expectations are also justified. The trajectories described in the previous section assume optimistic expectations that are justified in retrospect.

Our discussion suggests that in addition to an equilibrium in which all sectors end up adopting the new technology, there also may exist equilibria in which one or more of the industries continue to use the old technology and no one in these sectors invests in R&D. Another point that becomes clear from this discussion is that not only expectations about whether there will be enough product development in a particular sector matters; a potential investor has to form also expectations about the timing of product development by his rivals. If, for example, a company expects that all the others will start product development in the very distant future, and that as a result manufacturers of final output will switch to the new technology in a very distant point in time, then this company finds it unprofitable to invest today. As a result it delays product development.

Figure 4.6 presents simulated equilibrium trajectories for a two-sector economy in which the only difference between the sectors is the critical number of components that need to be developed in order to induce manufacturers of final output to switch to the new technology. In panel a the new GPT is first adopted by the industry that needs more components, while in panel b it is first adopted by the industry that needs fewer components. Importantly both are equilibrium trajectories on which investors have perfect foresight. They illustrate the point made above, that there typically exist multiple equilibria that differ in the order of adoption. This two-sector example has two additional equilibria: one in which only the sector with the smaller critical number n_{ic} develops components for the new GPT (and the other sector remains with the old technology), and one in which only the sector with the larger critical number n_{ic} develops components for the new GPT. Evidently multiplicity of equilibria is a generic property of this type of economies.

4.5.2 Efficiency

Which order of adoption is preferable? To answer this question, we need to compare the values of $U(0)$ (given in (12)) that result from alternative trajectories. Let $U^j(0)$ be the utility level that obtains from some trajectory j, which is characterized by a particular order of adoption. Then it follows from (7) and (12) that the difference in utility levels from trajectories j and k is given by

$$U^j(0) - U^k(0) = \int_0^{+\infty} e^{-\rho t}[\log E^j(t) - \log P^j(t)]\, dt$$

$$- \int_0^{+\infty} e^{-\rho t}[\log E^k(t) - \log P^k(t)]\, dt,$$

$(L=1; a_i=1; \alpha=0.5; \beta_i=0.5; \rho=0.02)$

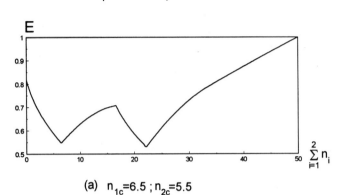

(a) $n_{1c}=6.5$; $n_{2c}=5.5$

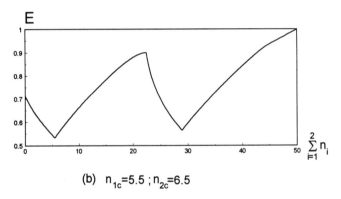

(b) $n_{1c}=5.5$; $n_{2c}=6.5$

Figure 4.6
Different orders of adoption

where $E^j(t)$ is the spending level at time t on trajectory j and $P^j(t)$ is the value of the price index (26) at time t on trajectory j. Once the suitable trajectories are known, we can calculate these welfare differences.

For the two trajectories depicted in figure 4.6, we find that welfare is higher in panel b, when the sector that needs fewer components in order to switch to the new technology invests first. This accords well with intuition because the sector with the larger critical value n_{ic} has a disadvantage over the other one, in the sense that it needs to develop more components in order to make the new technology viable. Therefore it is better to let it develop last when the price of consumption is low, and therefore when the alternative cost of R&D is lower.

Although this welfare ranking makes sense and we found it in all the simulations, it is important to note that it is not entirely trivial. To see

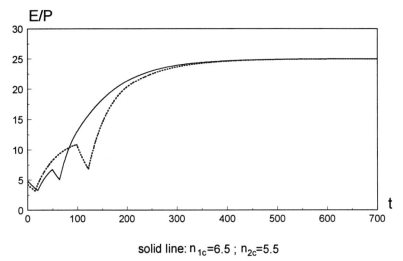

Figure 4.7
Welfare comparison

why, consider figure 4.7 where we plot real consumption E/P for trajectories with alternative orders of entry. Evidently, when the disadvantaged sector invests first real consumption is slightly higher for some time, lower for a span of time afterward, and it becomes higher again. An examination of the figure suggests that the welfare comparison may depend on the subjective discount rate. But we were not able to find a case in which early entry of the disadvantaged sectors is more desirable.

The association between welfare and the order of adoption is also clear-cut when sectors differ in spending shares. Intuitively one would expect higher welfare when the sector with the larger spending share invests first, and this has turned out to be the case in all the simulations. Finally we examined differences in labor inputs per unit R&D. According to intuition, welfare should be higher when the sector with the lower R&D costs invests first, and this is indeed confirmed by our simulations.

These examples demonstrate that the desired order of adoption depends in a clear way on the underlying parameters, when we examine one parameter at a time. When there are conflicting differences in more than one parameter, however, the desired order of adoption becomes a complicated function of the economy's structure. Thus, for example, if the sector with the larger spending share also requires more inputs per unit R&D, then it is not possible to provide a simple characterization of the required strength of these differences that tilts the welfare advantage in favor of one sector rather than the other. As a result there do not exist simple criteria upon which to judge which order of adoption is preferable.

Moreover, given that there are multiple equilibria, we may observe efficient as well as inefficient trajectories; there does not appear to exist a market mechanism that will coordinate expectations upon which the order of investment is determined so as to ensure the emergence of efficient outcomes. As a result, in a multisector economy many outcomes are equally plausible. In such circumstances there is plenty of room for historical accident and path dependence.

4.5.3 Limitations

Two limitations of the model need to be brought to the forefront. The first is that we assume that once a new GPT appears, its relative productivity level G_i remains constant over time. Of course the model does allow for further improvements, but only trivially, in the form of the appearance of a new GPT. This is clearly in sharp contrast to what we observe in reality: These technologies keep evolving and improving, within the span of what we would regard as a given GPT. Moreover the cumulative effect of these improvements can be very substantial, such that by the time a new GPT appears, it has to compete with a much better incumbent. We have not incorporated a changing GPT simply because the model is already too complex to handle this additional feature. It would seem though that a changing GPT would have two opposing effects. Within the span of a given GPT, diffusion and growth would certainly accelerate. However, and for the same underlying reasons, the adoption of new GPTs may be delayed.

The second limitation refers to the absence in the model of lateral linkages (or spillovers) between adopting sectors. In fact it is quite likely that sectors that are "close" to each other in the relevant metric[7] will benefit from positive externalities from neighboring sectors. For example, there

---

7. Such as "technological proximity" to the new GPT. An example would be the extent to which potential user sectors relied on vacuum tubes prior to the appearance of the transistor (see section 4.6).

may be a learning effect of the following form:

$$a_i = \phi_i(n_{i-1}), \qquad \phi' < 0.$$

That is, the more components for the new GPT developed by the "previous" sector, the lower the development costs of the sector next in line. This constitutes then a supply-push mechanism, which may play an important role in the diffusion process. We have explored extensions of the model that incorporate this feature, but in order to make it operative, we had to resort to other limiting (and equally unsatisfactory) assumptions, such as the absence of a "second wave." It would certainly be interesting to try to pursue extensions of the model in this direction.

4.6 "Reality Check"

In this section we sketch the initial diffusion stages of the paramount GPT of our times, semiconductors, and seek to characterize both the leading and the laggard user sectors, in terms of the model developed above.

In December 1947 John Barteen and Walter Brattain—of the Bell Labs—demonstrated the feasibility of the point contact transistor.[8] In January 1948 they applied for a patent, and in July of that year Bell Labs announced it to the world. A new GPT, semiconductors, was born. Almost half a century later we are still in the midst of the semiconductor era, which is leading us into the next stage in its meteoric rise, the so-called information age. One of the idiosyncratic features of this GPT is that it keeps getting smaller as it gets more powerful. Thus paradoxically, transistors, integrated circuits, microprocessors, and similar microelectronic components tend to be further removed from our sight and perception as they become ever more pervasive. It is therefore easy to overlook the fact that semiconductors are very much at the heart of the current wave of feverish innovation revolving around the Internet, and around the much heralded convergence of telecommunications and computers.

The first decade in the evolution of semiconductor technology was characterized by rapid advances in the manufacturing of transistors (which made them a viable commercial component), by the gradual switch from germanium to silicon, and by vast improvements in the performance

---

8. William Shockley was also a key player in the team that developed the transistor, and indeed the three of them received the Nobel Prize in Physics for their achievements in this field. In 1947–48 Shockley was pursuing a different track that proved to be superior to the initial point contact transistor: that of the junction transistor, which he demonstrated in 1951.

characteristics of the transistors themselves. The next quantum leap occurred in 1959 with the invention of the integrated circuit (by Jack Kilby, of Texas Instruments), which allowed for the packing of a large number of electronic components in a single tiny chip. A decade later it was the turn of VLSI (very large-scale integration), and later on that of the microprocessor, the heart of personal computers, developed at Intel.

Even though all of these developments are part and parcel of the same GPT, in terms of our model each can be thought of as a further upward jump in G_i;[9] that triggers new waves of complementary developments and hence new dynamic trajectories. We concentrate here on the first decade of the new GPT, that of the transistor, and explore which sectors were the first to adopt it and how they can be characterized in terms of the parameters of our model. Similarly we look into those sectors that were thought at the time as natural candidates for rapid adoption, but turned out in fact to be laggards. The early user sectors were hearing aids and computers.[10] The prominent laggards were telecommunications and automobiles.[11]

4.6.1 Early Adopters

Hearing Aids

The first sector to adopt the transistor almost immediately after it went into commercial production (in the early 1950s) was hearing aids. In fact by 1953 more than 15 manufacturers of hearing aids were buying transistors from Raytheon, then the leading producer of point contact transistors. What made the transistor so attractive for hearing aids at that early stage was first and foremost a large relative technological advantage vis-à-vis the incumbent technology, vacuum tubes (to be referred also as "valves").[12] First, transistors were much smaller and lighter than the (miniature) valves they replaced: The volume of a transistor-based hearing

---

9. Note that these further jumps have to be totally unexpected in order to qualify as such in the context of our model. Of course the G_i's keep advancing over time within these subperiods (or "generations"), but our model ignores that.
10. By "sector" we mean here a particular end-product or industry that uses semiconductors as components. It may be literally a whole sector (in the SIC sense), or more likely, a particular product family (e.g., computers).
11. The discussion here draws heavily from Braun and MacDonald (1982), hereforth just BM.
12. It is important to emphasize that what counts is indeed the relative technological lead in the context of the particular sector. That is, it may be that G_i is large for a given sector not because the new GPT (which is common to all sectors) is technologically very advanced at that stage but because the previous GPT is particularly inadequate for the needs of that sectors. This was clearly, and strikingly, the case for early computers (see below).

aid was about 45 cc versus 130 cc for a valve-based device.[13] Second, the power requirements were significantly lower, and hence the batteries that operated the system lasted much longer. In fact the battery life increased *8 times*, and the annual battery replacement cost went down from $40 to just $3. Third, transistors acted instantly, since there was no filament to heat. In terms of our model these relative advantages would be captured by a large G_i.

But there were also distinct disadvantages: Early transistors were noisier than valves, more restricted in their frequency performance, and more liable to damage by power surges and high temperatures. And there was a large price differential: Early transistors were much more expensive than valves (8 times more), and hence the price of transistor-based hearing aids was 3–4 times as high as conventional ones: $150–200 (in 1953) versus $50 for hearing aids using subminiature valves. Nevertheless, hearing-impaired people were willing to pay the much higher price, which implies that they put a high value on the relative advantages of the new technology. However, note that $200 was not an exorbitant price in 1953 (it would be $1,140 in 1995 dollars), and hence that this was a very *affordable* early application. Of course this was a very small market, made up of a well-defined group of users, that could easily evaluate the price/performance of a transistor-based system vis-à-vis the incumbent technology.

Computers

One of the most attractive initial uses of transistors was in early computers. Electronic computers were first built just before the transistor came into being (in the late 1940s). These were electronic monsters, containing thousands of valves, occupying large areas, requiring vast power consumption, and facing tremendous problems because of valve failure, overheating, and the like. The transistor was thus ideally positioned to replace the valves and bring about dramatic improvements in those dimensions as well as in speed (primarily because of the much shorter circuitry). Already in 1955 IBM brought out a transistor-based computer that replaced over 1,000 valves with 2,200 transistors; the new model was significantly smaller, there was little need for cooling, and power consumption was reduced by 95 percent. The series IBM 7000 quickly followed, using high-speed transistors (e.g., the IBM 7090 ran about 5 times as fast as the

---

13. It is interesting to note that valves had been getting smaller for over a decade prior to the arrival of the transistor, and that they were already surprisingly small. Still the transistor was an order of magnitude smaller.

valve-based 709; see Williams 1985). Transistors rapidly and completely displaced vacuum tubes in computers, and the computer sector established itself as the leading user of semiconductors (outside the military) for decades to come.[14]

As with hearing aids, early users displayed a high willingness to pay for the much faster, transistor-based computers, and there were relatively few of them. However, the underlying market could not have been more different. Most of the early computers were sold (or leased) to U.S. government agencies, particularly the military and defense contractors (having at their disposal generous budgets), and they cost several hundred thousand dollars.[15] The number of computers sold in the early 1950s was very small (a total of 250 computers by 1955), and overall demand was estimated at first to be very restricted.[16] However, dramatic advances in both hardware and software and concomitant reductions in "real" prices brought about an enormous expansion in the range of scientific and commercial applications, and hence in the size of the market. Computers are thus one of the first, and certainly the most striking example of the power of the new GPT, not just as a better substitute for older technologies but as the key to a whole range of new applications and uses.

Other Early Adopters

Two additional sectors have to be mentioned in this context: First, the military played an extremely important role almost from the beginning, both as the largest single user of transistors (since the mid-1950s) and in fostering research and innovation in this field. However, we focus here on *commercial* sectors, and hence we do not expand on it. Second, radios were also an important early adopter: Texas Instruments introduced the transistor radio in October 1954, in 1955 Raytheon brought out its own model (selling for $80), and shortly afterward car radios were also built with transistors. Transistor radios rapidly gained wide acceptance, and in fact they became the first transistor-based *mass* consumer product. The lack of more detailed information about this sector prevents us from further elaborating on it.

---

14. Computers accounted for 31% of all semiconductors outside the government (and hence outside the military) in 1963, 65% in 1978, and 55% in 1983 (BM, p. 80, and p. 144).

15. In 1955 the average price of CPUs of computer systems was over $600,000, and it represented about 80 percent of the total cost of the system (see Flamm 1987, table A.1). Nominal prices declined significantly over the course of the following decade (the average CPU cost $88,000 in 1965), and of course quality-adjusted prices dropped at a much faster rate.

16. A survey by the Department of Commerce conducted in the late 1940s estimated that about 100 mainframe computers would satisfy the entire needs of the nation (see BM, p. 69).

The following figures give a sense of the relative weight of the sectors reviewed above, as well as of other adopting sectors in the market for transistors, by the end of the decade under examination (adapted from Braun and McDonald 1982, p. 80, table 7.5):

Value of U.S. transistors by usage, 1963 ($M)

Military	119
Computers	42
Radios	33
Communications	16
Test and measuring	12
Controls	12
Other industries	12
Hearing aids and organs	7

4.6.2 Laggards[17]

Telecommunications
Writing in the late 1990s, it is ironic to talk about the telecommunication sector ("telecom" in short) as a laggard in the adoption of semiconductor technology. After all, we have seen in the past two decades the digitization of the telecom sector, the successful marriage of computers and telecommunications, and the proliferation of new modes of telecommunications based on both. Moreover these developments constitute one of the outmost expressions of the power of semiconductors, of their extremely wide reach, and of their potential for bringing forth deep and pervasive changes in the economy and society.

Yet the telecom sector turned out to be a laggard in the early era of the transistor: Adoption of semiconductor technology was extremely slow, and it did not have a significant impact on the sector as a whole until much later. This was so despite the fact that the drive toward the development of semiconductors at Bell Labs was motivated to a significant extent by the perception that the telecom sector would greatly benefit from the incipient technology. Consequently telecom was widely seen at the time as the natural candidate for early and widespread adoption of

---

17. Another prominent laggard was television sets. It seems that the dominant reason for the slow incorporation of transistors in TV sets was purely technological, but this requires further investigation.

the transistor. In that sense this is as good an example as any of wrong technological expectations.

The flip side of the potential for great benefits is the potential for extensive disruption. Telecom is not a single thing but a huge, intricate, and delicate system that poses strict demands on the technologies from which it feeds (compatibility, reliability, durability, etc.). Moreover technology is embedded in a massive stock of capital, and hence any change in the underlying GPT requires large capital outlays and concomitant adjustment costs. The sheer magnitude of these outlays, coupled with the rapid pace of change in semiconductor technology and the difficulties in forecasting future developments, made it extremely difficult to undertake decisions about adoption.[18] The result was that the adoption of semiconductor technology in the telecom sector proceeded at a much slower pace than anticipated. Summarizing the difficulties of adoption, Braun and MacDonald (1982) write:

It is because the adoption of this technological innovation affects so much else that rapid and widespread diffusion of the sort experienced with calculators and electronic watches is unlikely. No matter how great the benefits—and they are likely to be massive eventually—they would be exceeded by the costs of chaos occasioned by the immediate wholesale adoption of such revolutionary applications of semiconductor technology. (BM, p. 95)

Automobile Industry
Similarly to the case of the telecom sector, many observers saw the automobile as one of the greatest potential markets for semiconductor components already in the 1950s (BM, p. 200). There is room for the use of such components in a large range of safety devices, in engine control systems (to reduce gas consumption and pollution, and increase engine efficiency), in monitoring, instrumentation, and display, in servicing, and so on. However, this potential was extremely slow to materialize.

A number of technical reasons made it difficult to incorporate semiconductors in cars: First, the conditions inside a car engine are very unfavorable to the introduction of microelectronic components in that it is very hot, dirty, and noisy. Second, components in a car must function for years, and the car must be serviced by plain mechanics that had little knowledge of electronics. Third, while the prices of semiconductor components have declined steadily, that has not been the case for much of

---

18. See, for example, Braun and MacDonald's detailed account of the mistakes made by the British Post Office (now British Telecom) in the adoption of switching systems in the 1960s and 1970s (BM, pp. 196–198).

the sensor equipment needed to feed the information about the car's performance to the integrated circuits.

However, other sectors had faced even greater technical challenges and yet adopted semiconductors much earlier; indeed those difficulties fostered innovations that made adoption possible for a wider range of sectors. The real hindrance to early adoption in the automobile industry seems to lie elsewhere: Given the massive capital sunk in existing manufacturing plants, any significant change in the design of cars, or in the type of components used, requires extensive retooling and changes to the production line that are difficult and expensive to implement. It is not so much what semiconductors could do to the cars themselves but rather what they "threatened" to do to the production systems that became the decisive factor. Furthermore the automobile industry in the United States was at the time (the 1950s and 1960s) a tight, cozy oligopoly that resisted change and innovation in many dimensions, not just with respect to the new GPT. It was not until the late 1970s that the industry finally started to incorporate semiconductors in earnest, not only for mandated safety or pollution control features, and not just as optional features for upscale models, but for a wide range of monitoring and control functions in most models.[19]

4.6.3 Characterization of Early and Late Adopters

Recall that the key determinants of adoption in our model are $n_{ic} = n_i^o/(G_i)^{\alpha/(1-\alpha)}$ and β_i/a_i. The smaller the n_{ic}, the smaller the number of complementary components that have to be developed in order to adopt the new GPT, and hence the higher the likelihood of early adoption. The larger the β_i/a_i, the larger the expected demand for products using the new GPT relative to the development costs, making early adoption more likely. The discussion above suggests that the early adopters of the transistor can be characterized as follows:

1. Very large G_is, due not so much to advanced capabilities of the new GPT per se (early transistors were quite rudimentary) but rather to the obvious deficiencies of the previous GPT in those particular uses.

2. In the case of computers, it was also a low n_i^o. Electronic computers were born shortly before the transistor was invented, and hence there was not much time to perfect computers based on vacuum tubes.

---

19. In 1978 the estimated semiconductor component value of factory installations in automobiles stood at $111 million; by 1983 it had reached $620 million (BM, p. 202). The semiconductor contents of cars have kept increasing rapidly since then.

3. Although we do not have direct evidence to that effect, it would seem that in both sectors a_i was low. Since transistors were used in hearing aids and in computers primarily as substitutes for valves, incorporating them into the systems did not involve high development costs. It is when the introduction of semiconductors involve new functionalities, and/or the redesign of the systems in order to accommodate them, that a_i's are high.

4. High willingness to pay for the relative advantages that the new GPT offered in those uses, by a (initially) small but well-defined group of users. This translates into a (sufficiently) high and relatively certain stream of revenues that could at least cover production and development costs. Our model has just one demand parameter, β_i, and hence it cannot tell apart a high willingness to pay from, say, a high level of demand. Thus, we just associate these favorable demand conditions with a high manifest β_i.

It is very hard to form solid judgments on the role that each of these factors may have played, but two issues are worth noting. The first is that it may not be a coincidence that for early adopters virtually all four parameters were particularly advantageous; in fact, casual comparisons with other sectors that adopted later suggest that it is indeed rare to find sectors for which this is the case. If pressed to single out one factor as particularly prominent, it would seem that it was characteristic 1 above, that is, the plain inadequacy of the incumbent technology vis-à-vis the new if still rough GPT. And then of course there is historical accident, which in this case takes the form of a remarkable coincidence for hearing aids: As a memorial to Alexander Graham Bell, who took a personal interest in the fate of the deaf, Bell Labs waved royalties on the transistor patents to manufacturers of hearing aids. As to computers, the "accident" is in the timing, that is, the fact that they were developed just before the advent of the transistor and hence benefited (in retrospect) from a very low n_i^o.

As to the laggards, it seems quite clear that the decisive impediment to early adoption was, in terms of our model, an extremely large n_i^o. In the case of telecommunications it was just the sheer size of the system itself and the complex interdependence of subsystems and components. In the case of automobiles the n_i^o referred to established manufacturing plants and the disruption that the introduction of new components could bring on the production system. It is interesting to note that in this context, past investments in the old GPT are not "sunk," since they determine the relative efficiency of the old versus the new GPT. The larger those past investments are, the higher the efficiency threshold that lies in front of the new GPT in order to be adopted.

In retrospect it is clear that the errors of forecast made initially with respect to the laggards were due to the fact that these predictions were made primarily on the basis of the expected G_is (in that respect they were right for both the telecom and the auto sectors) and not on the basis of n_i^o, or on the "composite" measure n_{ic}. By identifying these factors, our model helps at the very least think systematically about this complex and yet crucial set of issues.

References

Braun, E., and S. MacDonald. 1982. *Revolution in Miniature: The History and Impact of Semiconductor Electronics*, 2d ed. Cambridge: Cambridge University Press.

Bresnahan, T., and M. Trajtenberg. 1995. General purpose technologies: "Engines of growth"? *Journal of Econometrics* 65: 83–108.

David, P. A. 1991. Computer and dynamo: The modern productivity paradox in a not-too-distant mirror. In *Technology and Productivity: The Challenge for Economic Policy*. Paris: OECD.

Flamm, K. 1987. *Targeting the Computer*. Washington: Brookings Institution.

Griliches, Z. 1957. Hybrid corn: An exploration in the economics of technological change. *Econometrica* 25: 501–22.

Mansfield, E. 1968. *Industrial Research and Technological Innovation*. New York: Norton.

Williams, M. R. 1985. *A History of Computing Technology*. Englewood Cliffs, NJ: Prentice Hall.

5 On the Macroeconomic Effects of Major Technological Change

Philippe Aghion and
Peter Howitt

5.1 Introduction

How can the adoption of new technological paradigms entail *cyclical* growth patterns including long recession periods? Among various attempts to address this question,[1] a particularly promising one is that of Helpman and Trajtenberg in chapter 3. The basic idea of the HT model is that general purpose technologies do not come ready to use off the shelf. Instead, each GPT requires new intermediate goods before it can be implemented. The development of these intermediate goods is costly, and the economy must wait until some critical mass of them has been developed before firms will switch from the previous GPT. During the waiting period, national income will fall as resources are taken out of production and put into R&D activities aimed at the development of the new intermediate components.

This chapter aims at clarifying GPTs' relationship to endogenous growth theory, simplifying its logical structure, pointing out empirical questions raised by the theory, and suggesting extensions. We first show how a simple version of the HT model can be derived from the basic Schumpeterian model of endogenous growth by adding a second stage to the innovation process, a stage of component-building. This simple version endogenizes the arrival times of successive GPTs, which HT take as exogenous. We then discuss some of the empirical shortcomings of this

---

Aghion: EBRD, London, and University College London. Howitt: The Ohio State University and CIAR. Special thanks are due to Paul David for helpful discussions.
1. Earlier contributions include Jovanovic and Rob (1990), who generated Schumpeterian waves based on the dichotomy between fundamental and secondary innovations, with each fundamental innovation being followed by a sequence of incremental innovations. Of particular interest as a macroeconomic model is the Cheng-Dinopoulos (1992) paper in which Schumpeterian waves obtain as a unique [non-steady-state] equilibrium solution, in which the current flow of monopoly profits follows a cyclical evolution.

two-stage theory and suggest ways to remedy them by adding a third stage, a stage of technology-spillover, during which firms learn how to adopt a new GPT not only by their own research but also by observing how other firms have successfully adopted it. We lay out a rudimentary version of such a three-stage model in which technology spillover is modeled as in the epidemiology literature, and present numerical simulations that illustrate how the approach might deal with some of the empirical shortcomings of the two-stage GPT model.

5.2 Basic Schumpeterian Growth Model

The simplest macroeconomic model of a GPT is the basic Schumpeterian growth model developed in Aghion-Howitt (1992). Final output is produced according to the production function:

$$y = AF(x),$$

where x is the flow of intermediate input, F is a production function with a positive and diminishing marginal product, and A is a productivity parameter measuring the quality of the intermediate input. Intermediate input itself is produced with labor according to a one-to-one linear technology so that x is also the flow of manufacturing labor. The supply of labor is an exogenous constant L.

Growth results from vertical innovations. Each innovation simultaneously invents a new method for producing final output throughout the economy (a new GPT) and a new intermediate good with which to implement the method. Thus it augments the current productivity parameter A by the multiplicative factor $\gamma > 1 : A_{t+1} = \gamma A_t$. Innovations in turn are the (random) outcome of research, and are assumed to arrive discretely with Poisson rate λn, where n is the current flow of labor used in research.

In steady state the allocation of labor between research and manufacturing remains constant over time and is determined by the arbitrage equation:

$$\omega = \lambda \gamma v, \qquad (A)$$

where the LHS of (A) is the productivity-adjusted wage rate $\omega = w/A$ that a worker earns by working in the manufacturing sector, and $\lambda \gamma v$ is the expected reward from investing one unit flow of labor in research. The productivity-adjusted value v of an innovation is determined by the Bellman equation:

$$rv = \tilde{\pi}(\omega) - \lambda n v,$$

where $\tilde{\pi}(\omega)$ denote the productivity-adjusted flow of monopoly profits accruing to a successful innovator and where the term $(-\lambda n v)$ corresponds to the capital loss involved in being replaced by a subsequent innovator.

The above arbitrage equation, which can be reexpressed as

$$\omega = \lambda \gamma \frac{\tilde{\pi}(\omega)}{r + \lambda n}, \tag{A}$$

together with the labor-market-clearing equation

$$\tilde{x}(\omega) + n = L, \tag{L}$$

where $\tilde{x}(\omega)$ is the manufacturing demand for labor,[2] jointly determines the steady-state amount of research n as a function of the parameters λ, γ, L, r. Figure 5.1 depicts the two curves (A) and (L), and shows the straightforward comparative statics results: $dn/d\lambda > 0$, $dn/d\gamma > 0$, $dn/dL > 0$, $dn/dr < 0$.

In steady-state the flow of consumption good (or final output) produced between the tth and the $(t+1)$th innovation is

$$y_t = A_t F(L - n),$$

which implies that in real time (when we denote by τ) the log of final output will increase by $\ln \gamma$ each time a new GPT arrives. Thus growth will be *uneven* (see figure 5.2), with the time path of $\log y(\tau)$ being a random step function. The average growth rate will equal the product of $\ln \gamma$, the size of each step, and λn, the average number of innovations per unit of time, that is, $g = \lambda n \ln \gamma$.

Although it fluctuates, the time path of aggregate output as depicted above does not involve a slump. Accounting for the existence of slumps requires another stage to be added to the innovation process as in the HT model, to which we now turn.

5.3 A Simple Helpman-Trajtenberg Model

As before, there are L workers who can engage either in production of existing intermediate goods or in research aimed at discovering new

---

2. Here we have

$$\tilde{x}(\omega) = \arg\max_{x} (\underbrace{F'(x)}_{p(x)} x - \omega x)$$

and

$$\tilde{\pi}(\omega) = \max(F'(x)x - \omega x).$$

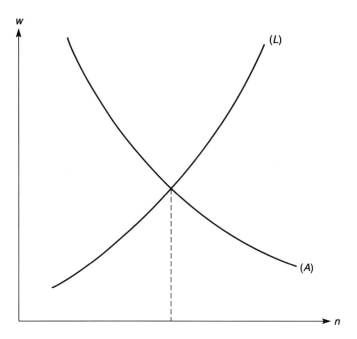

Figure 5.1
Steady-state equilibrium in the basic model of growth through creative destruction

intermediate goods. Again, each intermediate good is linked to a particular GPT. But now the discovery of a new generation of intermediate good comes in *two* stages. First a new GPT must come, and then the intermediate good must be invented that implements it.[3] Neither can come before the other. You need to see the GPT before knowing what sort of good will implement it, and people need to see the previous GPT in action before anyone can think of a new one. For simplicity we assume that no one directs R&D toward the discovery of a GPT. Instead, the discovery arrives as a serendipitous by-product of the collective experience of using the previous one.

Thus the economy will pass through a series of cycles, each having two phases, as indicated in figure 5.3. GPT_t arrives at time T_t. At that time the economy enters phase 1 of the tth cycle. During phase 1, the amount n of labor is devoted to research. Phase 2 begins at time $T_t + \Delta_t$ when this research discovers an intermediate good to implement GPT_t. During phase 2 all labor is allocated to manufacturing until GPT_{t+1} arrives, at

---

3. HT assume that some critical mass of intermediate goods must be invented before it can profitably be implemented, but we lose nothing essential by assuming that only one good is needed.

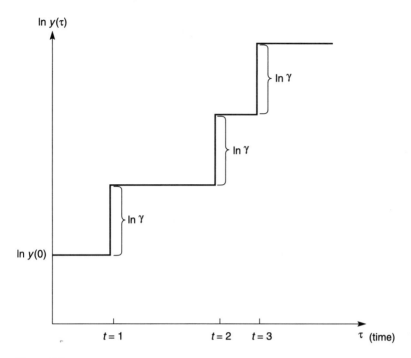

Figure 5.2
Log of GDP in steady-state equilibrium

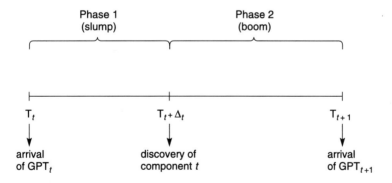

Figure 5.3
Two phases of GPT cycles

which time the next cycle begins. Over the cycle, output is equal to $A_{t-1}F(L-n)$ during phase 1 and $A_t F(L)$ during phase 2. Thus the drawing of labor out of manufacturing and into research causes output to fall each time a GPT is discovered, by an amount equal to $A_{t-1}[F(L) - F(L-n)]$.

A steady-state equilibrium is one in which people choose to do the same amount of research each time the economy is in phase 1, that is, where n is constant from one GPT to the next. As before, we can solve for the equilibrium value of n using a research arbitrage equation and a labor market equilibrium curve. Let ω_j be the wage and v_j the discounted expected value of the incumbent intermediate monopolist's profits when the economy is in phase j, each divided by the productivity parameter A of the GPT currently in use. In a steady state these productivity-adjusted variables will all be independent of which GPT is currently in use.

Since research is conducted in phase 1 but pays off when the economy enters into phase 2 with a productivity parameter raised by the factor γ, the usual arbitrage condition must hold in order for there to be a positive level of research in the economy:

$$\omega_1 = \lambda \gamma v_2. \tag{1}$$

Suppose that once we are in phase 2, the new GPT is delivered by a Poisson process with a constant arrival rate equal to μ. Then the value of v_2 is determined by the Bellman equation:

$$rv_2 = \tilde{\pi}(\omega_2) + \mu(v_1 - v_2). \tag{2}$$

By analogous reasoning, we have

$$rv_1 = \tilde{\pi}(\omega_1) - \lambda n v_1. \tag{3}$$

Combining (1) through (3) yields the research arbitrage equation:

$$\omega_1 = \frac{\lambda \gamma [\tilde{\pi}(\omega_2) + \mu \tilde{\pi}(\omega_1)/(r + \lambda n)]}{r + \mu}. \tag{4}$$

Since no one does research in phase 2, we know that the value of ω_2 is determined independently of research, by the market-clearing condition $L = \tilde{x}(\omega_2)$. Thus we can take this value as given and regard equation (4) as determining ω_1 as a function of n. The value of n is determined, as usual, by this equation together with the labor-market equation

$$L - n = \tilde{x}(\omega_1). \tag{5}$$

As in the basic model, the level of research n is an increasing function of the productivity of research λ, the size of improvement created by each

GPT γ, and the population L, and it is a decreasing function of the rate of interest r. The arrival rate μ of GPTs has a negative effect on research;[4] intuitively, an increase in μ discourages research by reducing the expected duration of the first of the two phases over which the successful researcher can capitalize the rents from an innovation. The size of the slump $\ln(F(L)) - \ln(F(L-n))$ is an increasing function of n, and hence it will tend to be positively correlated with the average growth rate.

The average growth rate will be the frequency of innovations times the size $\ln \gamma$, for exactly the same reason as in the basic model. The frequency, however, is determined a little differently than before because the economy must pass through *two* phases. An innovation is implemented each time a full cycle is completed. The frequency with which this happens is the inverse of the expected length of a complete cycle. This in turn is just the expected length of phase 1 plus the expected length of phase 2:

$$\frac{1}{\lambda n} + \frac{1}{\mu} = \frac{\mu + \lambda n}{\mu \lambda n}.$$

Thus we have the growth equation

$$g = \ln \gamma \frac{\mu \lambda n}{\mu + \lambda n}. \tag{6}$$

The expected growth rate will be positively affected by anything that raises research, with the possible exception of a fall in μ. In the limit, when μ falls to zero, growth must also fall to zero as the economy will spend an infinitely long time in phase 2, without growing. Thus for small enough values of μ, g and n will be affected in opposite directions by a change in μ.

One further property of this cycle worth mentioning is that as HT point out, the wage rate will rise when the economy goes into a slump. That is, since there is no research in phase 2, the normalized wage must be low enough to provide employment for all L workers in the manufacturing sector, whereas with the arrival of the new GPT, the wage must rise to induce manufacturers to release workers into research.

---

4. To show this, it suffices to show that an increase in μ shifts the research arbitrage curve to the left. By applying the implicit function theorem to (4), we see that the sign of this shift is

$$\operatorname{sgn} \frac{dn}{d\mu} = -\operatorname{sgn}\left(\omega_1 - \frac{\lambda \gamma \tilde{\pi}(\omega_1)}{r + \lambda n}\right).$$

Since no research is done in phase 2, labor market equilibrium requires that $\omega_2 < \omega_1$, and hence $\tilde{\pi}(\omega_2) > \tilde{\pi}(\omega_1)$. Applying this to (4) yields

$$\omega_1 > \frac{\lambda \gamma \tilde{\pi}(\omega_1)}{r + \lambda n} \frac{r + \lambda n + \mu}{r + \mu} > \frac{\lambda \gamma \tilde{\pi}(\omega_1)}{r + \lambda n}.$$

5.4 Empirical Questions Raised by the Theory of GPTs

There are two aspects of this theory that may call its empirical relevance into question. The first is the likely *size of the slump* that it might cause. All of the decline in output is attributable to the transfer of labor out of manufacturing and into R&D. But since the total amount of R&D labor on average is only about two and a half percent of the labor force in the U.S. economy, it is hard to see how this can account for a change in aggregate production of more than a fraction of a percent. The size of the slump would be even smaller if we assumed, as HT do, that some national income is imputed to the R&D workers even before their research pays off in a positive stream of profits in the intermediate sector.[5]

The second questionable aspect of this theory has to do with the *timing* of slowdowns: The HT model implies an *immediate* slump as soon as the GPT arrives. In fact, as David (1990) argues, it may take several decades before major technological innovations can have a significant impact on macroeconomic activity (David talks about a preparadigm phase of twenty-five years in the case of the electric dynamo). It is hard to believe that labor could be diverted from current production on a large scale into an activity that will pay off only in the very distant future.

The first of these problems is relatively easy to deal with (at least conceptually), for one can think of a number of reasons why the adjustment to a GPT would cause adjustment and coordination problems resulting in a slump. For example, as Atkeson and Kehoe (1993) analyze, the arrival of a new GPT might induce firms to engage in *risky experimentation* on a large scale with start-up firms, not all of which will succeed. The capital sunk into these start-up firms will not yield a competitive return right away except by chance; meanwhile national income will drop as a result of that capital not being used in less risky ways using the old GPT. Also an increase in the pace of innovation aimed at exploiting the new GPT may result in an increased rate of job turnover, and hence in an increased rate of unemployment (section 5.6.2). Greenwood and Yorukoglu (1996) present an analysis in which the costs of learning to use equipment embodying the new GPT can account for a prolonged productivity showdown. Howitt (chapter 9) shows how the arrival of a new GPT can cause output growth to slow down because it accelerates the rate of obsolescence of existing physical and human capital (see also section 5.6.3).

---

5. HT find that a measured slump occurs when the GPT arrives even if the full cost of R&D is imputed as national income. The reason is that the discovery induces workers to leave a sector where they produce rents (because the intermediate sector is imperfectly competitive) and to enter a competitive sector—research—where there are no rents.

The second problem is more challenging to deal with. The question is, if the exploitation of a new GPT is spread out over a period of many decades, why should it not result in simply a slow enhancement in aggregate productivity as one industry after another learns to use the new technology?

Again, several answers come to mind, and we actually think of the following three explanations as being complementary. First are *measurement* problems: As already stressed by David and others, it may take a while before the new products and services embodying the new GPT can be fully accounted for by the conventional statistics. (This, however, does not explain the possibility of delayed *slumps*.) Second, the existence of *strategic complementarities* in the adoption of new GPTs by the various sectors of the economy may generate temporary lock-in effects, of a kind similar to the implementation cycles in Shleifer (1986). It may then take a long time before a critical number of sectors decide to jump on the bandwagon of the new GPT.

A third explanation, which will be the main focus of our analysis in section 5.5, lies in the phenomenon of *technology-spillover*, which is a special case of *social learning*. That is, the way a firm typically learns to use a new technology is not to discover everything on its own but to learn from the experience of other firms in a similar situation, namely other firms for whom the problems that must be solved before the technology can successfully be implemented bear enough resemblance to the problems that must be solved in this firm. The procedures of those successful firms can then be used as a "template" on which to prepare for adoption in this firm.

Thus at first the fact that no one knows how to exploit a new GPT means that almost nothing happens in the aggregate. Only minor improvements in knowledge take place for a long time because successful implementation in any sector requires firms to make independent discoveries with little guidance from the successful experience of others. But, if this activity continues for long enough, a point will eventually be reached when almost everyone can see enough other firms using the new technology to make it worth their while experimenting with it. Thus, even though the spread of a new GPT takes place over a long period of time, most of the costly experimentation through which the spread takes place may be concentrated over a relatively short subperiod, during which time there will be is a cascade or snowball effect resulting sometimes in a (delayed) aggregate slump.

5.5 Technology-Spillover and GPTs

This section presents a model of the spread of technology, which is similar to the sorts of models used by epidemiologists when studying the spread of disease,[6] which also takes place through a process of social interaction between those who have and those who have not yet been exposed to the new phenomenon. The setting is like that of the model we have just described, with a continuum of sectors uniformly distributed on the unit interval, except that now each sector must invent its own intermediate good in order to exploit the GPT.

The innovation process in this model involves *three* stages. First the GPT is discovered. Then each sector discovers a "template" on which research can be based, as discussed in the previous section. Finally, that sector implements the GPT when its research results in a successful innovation. This final stage corresponds to the component-building stage of the HT model. The second stage is the new technology-spillover stage.

5.5.1 Basic Setup

We study here the nature of the cycle caused by the arrival of a single GPT, under the assumption that the arrival rate μ is so small that there is insignificant probability that the next GPT will arrive before almost all sectors have adopted the one that has just arrived. In order to simplify the analysis even further, we suppose that the amount of research in each sector is given by a fixed endowment of specialized research labor. Thus all the dynamics will result from the effects of social learning on the payoff rate to experimentation. This is the phenomenon that we believe to be at the heart of the timing of the delayed cyclical response to GPTs.[7]

Aggregate output at any point in time is produced by labor according to the constant returns technology:

$$Y = \left\{ \int_0^1 A(i)^\alpha x(i)^\alpha \, di \right\}^{1/\alpha}, \tag{7}$$

where $A(i) = 1$ in sectors where the old GPT is still used, and

---

6. See, for example, Anderson and May (1992).
7. A fuller analysis would endogenize the allocation of labor between research and manufacturing as in the one-stage and two-stage models described above. This would presumably accentuate the effects we find, since it would draw more labor into research, hence augmenting the intensity of experimentation just when the informational cascade we focus on is already having the same effect.

$A(i) = \gamma > 1$ in sectors that have successfully innovated, while $x(i)$ is manufacturing labor used to produce the intermediate good in sector i.

All sectors are in one of three states. In state 0 are those sectors who have not yet acquired a template. In state 1 are those who have a template but have not yet discovered how to implement it. In state 2 are those sectors that have succeeded in making the transition to the new GPT. We let the fraction of sectors in each state be represented by n_0, n_1, n_2, and suppose that initially $n_0 = 1, n_1 = n_2 = 0$.

A sector moves from state 0 to state 1 when a firm in that sector either makes an independent discovery of a template or discovers one by "imitation," that is by observing at least k "similarly located" firms that have made a successful transition to the new GPT (firms in state 2). The Poisson arrival rate of independent discoveries to such a sector is $\lambda_0 \ll 1$. The Poisson arrival rate of opportunities to observe m similarly located firms is assumed to equal unity. The probability that such an observation will pay off (in other words the probability that at least k among the m similar firms will have successfully adopted the new GPT) is given by the cumulative binomial:

$$\varphi(m, k, n_2) = \sum_{j=k}^{m} \binom{m}{j} n_2^j (1 - n_2)^{m-j},$$

since n_2 is the probability that a randomly selected firm will be in state 2. Thus the flow of sectors from state 0 to state 1 will be n_0 times the flow probability of each sector making the transition: $\lambda_0 + \varphi(m, k, n_2)$.

For a sector to move from state 1 to state 2, the firm with the template must employ at least N units of labor per period (the equivalent of n in the HT model). We can think of this labor as being used in formal R&D, informal R&D, or an experimental start-up firm. In any case it is not producing current output. Instead, it is allowing the sector access to a Poisson process that will deliver a workable implementation of the new GPT with an arrival rate of λ_1. Thus the flow of sectors from states 1 to 2 will be the number of sectors in state 1, n_1 times the success rate per sector per unit of time λ_1.

We can summarize the discussion to this point by observing that the evolution over time of the two variables n_1 and n_2 is given by the autonomous system of ordinary differential equations:

$$\begin{aligned} \dot{n}_1 &= [\lambda_0 + \varphi(m, k, n_2)](1 - n_1 - n_2) - \lambda_1 n_1, \\ \dot{n}_2 &= \lambda_1 n_1, \end{aligned} \tag{S}$$

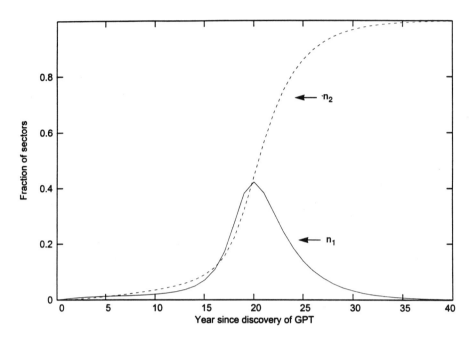

Figure 5.4
Computed paths of the fraction of sectors experimenting with the new GPT (n_1) and the fraction using the new GPT (n_2).

with initial condition $n_1(0) = 0$, $n_2(0) = 0$. (The time path of n_0 is then given automatically by the identity $n_0 \equiv 1 - n_1 - n_2$.)

Figure 5.4 depicts a numerical solution to the above system (S). Not surprisingly, the time path of n_2 follows a logistic curve, accelerating at first and slowing down as n_2 approaches 1, with the maximal growth rate occurring somewhere in the middle. Likewise the path of n_1 must peak somewhere in the middle of the transition, since it starts and ends at zero. If the arrival rate λ_0 of independent discoveries is very small, then both n_1 and n_2 will remain near zero for a long time. In the case shown in figure 5.4, $\lambda_0 = 0.005$, $\lambda_1 = 0.3$, $m = 10$, and $k = 3$. These will be the baseline parameter values used in the simulations below. The number of sectors engaging in experimentation peaks sharply in year 20 due to social learning.

The solution to the system (S) can be used with the aggregate production function (7), and the market-clearing condition for labor to determine the time path of aggregate output. Using the symmetry of the production function, which implies that all the sectors using the same GPT (either old or new) will demand the same amount of manufacturing labor, we can reexpress the flow of aggregate output as

$$Y = \left\{ \int_0^{1-n_2} x_0(i)^\alpha \, di + \gamma^\alpha \int_{1-n_2}^1 x_N(i)^\alpha \, di \right\}^{1/\alpha}, \tag{8}$$

where x_0 (resp. x_N) denotes the labor demand by a sector using the old (resp. new) GPT.

In this Cobb-Douglas world the local monopolists in sectors in state 0 and 1, who use the old technology, will demand labor according to the demand function[8]

$$x_0 = \left(\frac{w}{\alpha}\right)^{1/(\alpha-1)} Y, \tag{9}$$

while those in sectors in state 2 will demand according to

$$x_N = \left(\frac{w}{\alpha \gamma^\alpha}\right)^{1/(\alpha-1)} Y, \tag{10}$$

where w is the real wage rate.

Using (9), (10), and the market-clearing condition,

$$\underbrace{(1-n_2)x_0 + n_2 x_N}_{\text{manufacturing labor demand}} + \underbrace{n_1 N}_{\text{experimenting labor}} = L, \tag{L}$$

we can solve for the real wage w as a function of Y, n_1, and n_2. Substituting this solution back into the above expressions for x_0 and x_N and then substituting the resulting values of x_0 and x_N into (8) yields the following reduced-form expression for output:

$$Y = (L - n_1 N)(1 - n_2 + n_2 \gamma^{\alpha/(1-\alpha)})^{(1-\alpha)/\alpha}. \tag{11}$$

Figure 5.5 shows the time path of output that results from the above dynamics in n_1 and n_2 in the baseline case where $N = 6$, $L = 10$, and $\alpha = 0.5$. As expected, output is not much affected by the new GPT for the first decade and half, but then it enters a severe recession when the number of sectors engaging in experimentation increases sharply as a

---

8. This follows from profit-maximization: For any sector i, $x(i) = \arg\max_x \{p_i(x)x - wx\}$, where

$$p_i(x) = \frac{\partial Y}{\partial x} = \begin{cases} x^{\alpha-1} Y^{1-\alpha} & \text{if sector } i \text{ uses the } old \text{ technology} \\ \gamma^\alpha x^{\alpha-1} Y^{1-\alpha} & \text{if sector } i \text{ uses the } new \text{ technology.} \end{cases}$$

The corresponding first-order conditions, respectively, for old and new sectors yield the above equations (9) and (10).

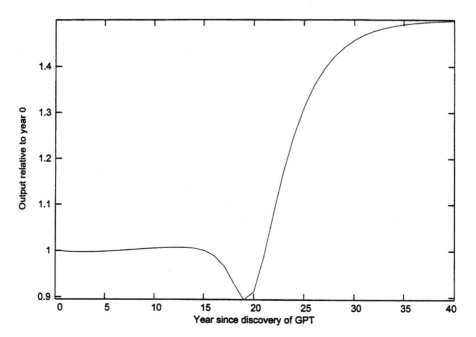

Figure 5.5
Computed path of output following the introduction of a new GPT

result of social learning: Output reaches a trough in year 19, after a 10.5 percent drop. From there it begins to grow, ultimately attaining a value of $\gamma(=1.5)$ times its original value.

The delay in the slump caused by the GPT could not have occurred without the impact of social learning.[9] That is, suppose that the function $\varphi(m, k, n_2)$ which embodies the effects of social learning were replaced by the constant $\varphi_0 = \int_0^1 \varphi(m, k, n_2)\, dn_2$ so that the average rate of learning would be the same but learning would not be affected by the process of observing other sectors that have succeeded in implementing the new GPT. Then n_2 would still follow a mild logistic curve, but the intensity of experimentation n_1 would rise immediately upon the arrival of the GPT and would fall monotonically from then on. Output could go through a slump, but the maximal rate of decline would occur immediately at year 0. The baseline case with no social learning is illustrated in figure 5.6.

Intuitively the reason why the slump cannot be delayed in this case is as follows: In order for output to be falling, there must be a positive flow of sectors into state 1, which is drawing workers out of manufacturing.

---

9. Greenwood and Yorukoglu (1996) assume a private learning process that also produces diffusion according to a mild logistic curve.

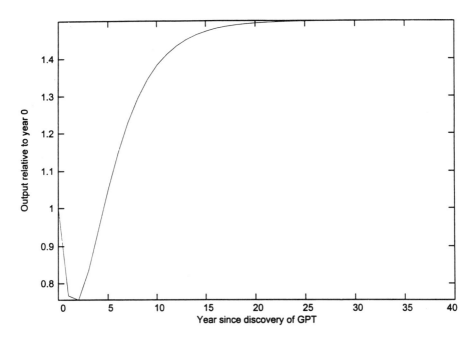

Figure 5.6
Computed path of GDP when there is no social learning

But without social learning this flow must be diminishing whenever a slump is underway, since the rise in the level of n_1 (and in n_2) will be reducing the pool $(1 - n_1 - n_2)$ from which new experimenters are drawn, while the rise in n_1 increases the flow of successful innovators out of the state of experimentation. (See the equation for $\dot{n}_1$ in (S) above.) Thus either the slump starts right away, in which case its intensity will diminish steadily, or it never starts at all.[10] What social learning does is to reverse

---

10. To see this more formally, suppose that the output function (11) can be approximated by its first-order Taylor expansion around $n_1 = n_2 = 0$:

$$Y \simeq L - N n_1 + \xi n_2, \quad \xi \equiv \frac{1-\alpha}{\alpha}(\gamma^{\alpha/(1-\alpha)} - 1)L > 0.$$

Because $\dot{n}_2$ is always positive, $\dot{n}_1$ must also be positive whenever Y is not rising. Thus if $\dot{Y} \leq 0$, then

$$\dot{Y} \simeq -N\dot{n}_1 + \xi \dot{n}_2,$$
$$\dot{Y} \simeq -N[(\lambda_0 + \varphi_0)(1 - n_1 - n_2) - \lambda_1 n_1] + \xi \lambda_1 n_1,$$
$$\ddot{Y} \simeq [N(\lambda_0 + \varphi_0 + \lambda_1) + \xi \lambda_1]\dot{n}_1 + N(\lambda_0 + \varphi_0)\dot{n}_2,$$
$$\ddot{Y} > 0 \quad \text{(because } \dot{n}_1 > 0, \dot{n}_2 > 0\text{)}.$$

Hence a delayed slump, with $\dot{Y}$ turning negative or becoming more negative at some date $t > 0$, is impossible.

Table 5.1
Sensitivity analysis

Parameter	Value	Slump	Size (%)	Peak date	Trough date
Baseline		Yes	11	12	19
α	0.2	Yes	12	12	19
(0.5)	0.8	Yes	8	15	19
γ	1.1	Yes	22	0	20
(1.5)	3.0	No			
k	1	Yes	23	0	5
(3)	5	Yes	4	37	42
m	3	No			
(10)	30	Yes	22	4	10
N	2	No			
(6)	8	Yes	20	11	19
λ_0	0.001	Yes	11	32	40
(0.005)	0.025	Yes	10	5	10
λ_1	0.1	Yes	22	13	29
(0.3)	1.0	No			

the effect of n_2 on the rate of growth of n_1; that is, as n_2 rises, the resulting increase in the likelihood of imitation counterbalances the fall in the number of possible imitators, thus causing the cascade at the heart of our analysis.

5.5.2 Some Comparative Dynamics

Table 5.1 shows how the time path of aggregate output responds to variations in the basic parameters of the model, namely

- α, which measures the degree of substitutability across intermediate inputs
- γ, which measures the size of productivity improvements brought about by the new GPT
- N, the number of workers taken out of manufacturing by each experimenting firm
- m, the number of sectors potentially "similar" to a given sector
- k, the required number of observations of successful experimentations in order to acquire a template "by imitation"
- λ_0, the arrival rate of independent ideas for new templates
- λ_1, the arrival rate of success for experimenting firms

In all cases the simulation produces either a marked slump, as in figure 5.5, or a monotonic increase in output. When there is no slump, there is an initial period of relatively slow growth followed by a sharp acceleration coming just after the peak in experimentation. When there is a slump, it almost always comes after a period of mild growth, which itself is often preceded by a very mild (less than half a percentage point) recession. The size of slump reported in table 5.1 is the percentage shortfall from the peak attained at the end of the period of mild growth (or from year 0 if no such period exits) to the trough. From table 5.1 we can make some observations:

1. The magnitude of slumps increases as α decreases, that is, when intermediate inputs become less substitutable. This is fairly intuitive: As α decreases, the downsizing of old manufacturing sectors, which results from labor being diverted away into experimentation, is less and less substituted for by the new—more productive—intermediate good sectors.

2. The magnitude of slumps decreases as γ increases, and for sufficiently large γ, the slump disappears. Again, this result is intuitive: The bigger productivity of new sectors compensates for the reduction in output in old sectors caused by experimentation (and by the resulting wage increase), thereby reducing the scope for aggregate slumps.

3. If m is too small, output grows steadily: Indeed the lower the m, the lower is the scope for social learning and for the resulting snowball effect on aggregate output.

4. An increase in k leads to bigger delays but smaller slumps: As k increases, it will take longer for "imitation" and social learning to become operational. Thus by the time a cascade begins, a higher number of sectors will have already moved into using the new and more productive GPT, and hence the smaller the size of aggregate slumps.

5. An increase in N leads to larger slumps. This is straightforward: The bigger the N, the more labor will be diverted away from manufacturing into experimentation by firms in state 1, and therefore the bigger the size of slumps when social learning causes the fraction of experimenting sectors to increase sharply.

6. An increase in the arrival rate of independent ideas λ_0 speeds up the macroeconomic response to the new GPT. This is not surprising, for the larger the λ_0, the faster the conditions will be created for social learning to operate.

7. An increase in the success rate of experimentation λ_1 reduces the size of slumps: This is again easy to understand, for the larger the λ_1, the faster

emerge the sectors using the new GPT which compensates for the downsizing of manufacturing activities induced by experimentation.

5.6 Accounting for the Size of Slowdowns

In this final section we sketch some extensions to the three-stage model of the previous section and show, by means of simulations, how these extensions might address the other question raised in section 5.4, namely that of how the slowdown caused by a new GPT could be larger than indicated by the relatively small fraction of the labor force engaged in formal R&D.

5.6.1 Skill Differentials

The last five years or so have witnessed an upsurge of empirical papers on skill differentials and wage inequality, and their relationship with technological change (in particular, see Juhn et al. 1993). It turns out that a straightforward extensions of our three-stage GPT model can account for an increasing skill differential as a result of the introduction of new GPT. The same extension can also magnify the slump.[11]

More formally, suppose that the labor force L is now divided into skilled and unskilled workers and that the implementation of the new GPT requires *skilled* labor, whereas old sectors can indifferently use skilled or unskilled workers to manufacture their intermediate inputs. Also let us assume that the fraction of skilled workers is increasing over time, such as a result of schooling and/or training investments which we do not model here:

$$L_s(t) = L(1 - (1-\tau)e^{-\lambda_2 t}), \qquad 0 < \tau < 1,$$

where τ is the initial fraction of skilled workers and λ_2 is a positive number measuring the speed of skill acquisition.

The transition from the old to the new GPT can then be divided into two subperiods. In the first subperiod the number of sectors using the new GPT is too small to absorb the whole skilled labor force, which in turn implies that a positive fraction of skilled workers will have to be employed by the old sectors at the same wage as their unskilled peers.

---

11. Our explanation of both the differential and the slowdown is similar in spirit to that of Greenwood and Yorukoglu (1996) who also emphasize the role of skilled labor in implementing new technologies.

Thus during the early phase of transition the labor market will remain "unsegmented," with aggregated output and the real wage being determined exactly as before.[12]

In the second subperiod the fraction of new sectors has grown sufficiently large that it can absorb the whole skilled labor force. The labor market therefore becomes segmented, with skilled workers being exclusively employed (at a higher wage) by new sectors, while unskilled workers remain in old sectors. Let w_u and w_s denote the real wages, respectively, paid to unskilled and skilled workers. The demand for manufacturing labor by the old and new sectors are still given by

$$x_0 = \left(\frac{w_u}{\alpha}\right)^{1/(\alpha-1)} Y,$$

and

$$x_N = \left(\frac{w_s}{\alpha\gamma^\alpha}\right)^{1/(\alpha-1)} Y,$$

except that we now have: $w_s > w_u$, where the two real wages are determined by two separate labor market clearing conditions, respectively:

$$L_s = n_1 N + n_2 x_N \tag{12}$$

and

$$L_u = L - L_s = (1 - n_2)x_0. \tag{13}$$

These equations yield

$$w_s = \gamma^\alpha \alpha \left(\frac{n_2 Y}{L_s - n_1 N}\right)^{1-\alpha}$$

and

$$w_u = \alpha \left(\frac{(1-n_2)Y}{L - L_s}\right)^{1-\alpha}.$$

Substituting for x_0 and x_N from (12) and (13) into (8) yields the following expression for aggregate output during the segmented phase of transition:

$$Y = [(1-n_2)^{1-\alpha}(L - L_s)^\alpha + n_2^{1-\alpha}\gamma^\alpha(L_s - n_1 N)^\alpha]^{1/\alpha}.$$

---

12. That is, by equations (L) and (9)–(11), we have $w = \alpha[1 - n_2 + n_2\gamma^{\alpha/(1-\alpha)}]^{(1-\alpha)/\alpha}$.

The cut-off date t_0 between the unsegmented and segmented phases of transition to the new GPT is determined by $w_s(t_0) = w_u(t_0)$.

Panel a of figure 5.7 depicts the time-path of real wages and panel b the time path of aggregate output in the baseline case of the previous section with $\lambda_2 = 0.05$ and $\tau = 0.25$. Two interesting conclusions emerge from this simulation:

1. The skill premium (w_s/w_u) starts increasing sharply in a year ($t = 21$) when social learning is accelerating the flow of new sectors in the economy, and then the premium keeps on increasing although more slowly during the remaining part of the transition process.[13] Since everyone ends up earning the same (skilled) wage, standard measures of wage inequality first rise and then fall.

2. Compared to the baseline case *without* skill differentials and labor market segmentation, the magnitude of the slump is the same (11 percent), but the recovery is slower. The reason is that high productivity sectors are constrained by the short supply of skilled labor; in simulations with other parameter values we see that the slump is exacerbated by the skill shortage if the market becomes segmented earlier.

5.6.2 *Job Search*

We now extend the basic setup in another direction by introducing costly job search, which, together with the destruction of jobs by new sectors, generates unemployment on the transition path. Unemployment in turn diverts a higher fraction of the labor force out of manufacturing, thereby *increasing* the size of slumps. Indeed slumps can now occur even if the labor N needed to perform experiments is negligible, as would be suggested by the small fraction of workers engaged in formal R&D in the U.S. economy.

More formally, suppose that the fraction β of workers in each sector that adopts the GPT (and moves in to n_2) goes into temporary unemployment because these workers are unable to adapt to working with the new GPT in the sector where they were formally unemployed. Suppose also that the fraction λ_3 of the unemployed per period succeeds in finding

---

13. The acceleration in the premium, with w_s increasing and w_u decreasing sharply at the beginning of the segmented phase, has to do with the high demand for skilled experimentation labor during the peak time of social learning. The skilled wage w_s starts tapering off thereafter because most sectors are already in phase 2 and the supply of skilled labor keeps on increasing.

On the Macroeconomic Effects of Major Technological Change 141

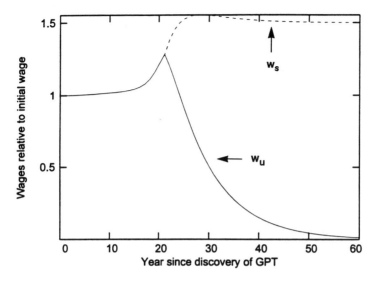

a

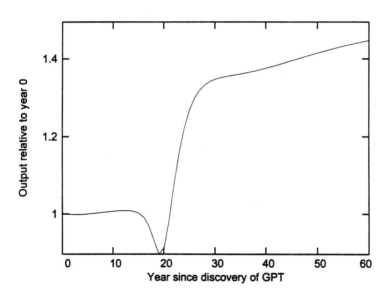

b

Figure 5.7
Skilled and unskilled wages, and GDP, when not everyone can work with the new GPT at first

a new job. Then the evolution of U, the number unemployed, is governed by

$$\dot{U} = \underset{\text{job destruction}}{\beta x_0(w)\lambda_1 n_1} - \underset{\text{job creation}}{\lambda_3 U}.$$

Output and the real wage are determined exactly as in the basic model of section 5.5 except with the "effective labor force" $L - U$ instead of L.[14] Putting this real wage into the demand function (9) and substituting for Y using (11) yields the equilibrium quantity:

$$x_0(w) = \frac{L - U - n_1 N}{1 - n_2 + n_2 \gamma^{\alpha/(1-\alpha)}}.$$

Figure 5.8 depicts the time paths of unemployment and aggregate output with the baseline parameter set from section 5.5 together with $\beta = 0.5$ and $\lambda_3 = 2$. The unemployment rate reaches a sharp peak in year 20, just after experimentation reaches its peak, with the predictable effect of increasing the size of the slump (from 11 to 13 percent).

5.6.3 Obsolescence

There is another, and maybe more straightforward, explanation for the slowdowns or slumps induced by major technological changes, one that should occur immediately to anyone familiar with Schumpeter's ideas: namely the (capital) obsolescence caused by the new wave of (secondary) innovations initiated by a new GPT. To capture this idea, we first reinterpret the model of section 5.5 by supposing that the factor used in both production and research is not labor but capital, either physical or human. Each time an innovation arrives implementing the GPT in a sector, it destroys a fraction δ of the capital that had previously been employed in that sector because all capital must be tailor-made to use a specific technology in a specific sector, and some of the capital is lost when it is converted to use in another sector or with another technology.[15]

For simplicity, suppose that people are target savers; that is, they save a constant fraction s per period of the gap between the desired capital stock

---

14. For simplicity, we identify flows into unemployment with flows out of the labor force. This allows us to bypass the technical complications involved in modeling explicitly the bargaining game between new sectors and workers. Taking the latter more traditional modeling route would needlessly complicate the algebra.
15. This is the assumption made in Howitt in chapter 9.

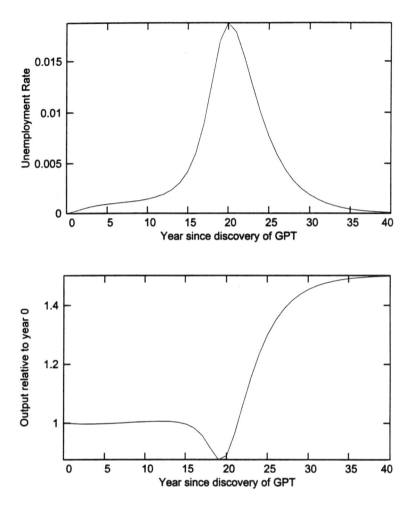

Figure 5.8
Unemployment and GDP when it takes time to form labor-market matches

L and the actual stock K. Then the rate of net accumulation of capital is

$$\dot{K} = s(L - K) - \delta x_0(w)\lambda_1 n_1.$$
$\phantom{\dot{K} = }$ gross saving obsolescence

Output and the real wage (the real rate of return to capital) are determined as in the basic model of section 5.5, but with L replaced by K. The initial stationary state with $n_1 = 0$ has $K = L$.

It is easy to see that this modification of the basic model of section 5.5 is formally equivalent to that of the previous section, with the gap $L - K$

replacing the number unemployed U, the saving rate s replacing the job-finding rate λ_3, and the obsolescence fraction δ replacing the job-destruction fraction β. Thus, for the same reasons as in the previous section, the capital shortfall will peak sharply around the same time as the peak in experimentation, and the slump will be larger than if there were no obsolescence.

References

Aghion, P., and P. Howitt. 1992. A model of growth through creative destruction *Econometrica* 60: 323–51.

Anderson, R. M., and R. M. May. 1992. *Infectious Diseases of Humans: Dynamics and Control.* Oxford: Oxford University Press.

Atkeson, A., and P. Kehoe. 1993. Industry evolution and transition: The role of information capital. Unpublished. University of Pennsylvania.

Cheng, L., and E. Dinopoulos. 1992. A Schumpeterian model of economic growth and fluctuations. Mimeo. University of Florida.

David, P. 1990. The dynamo and the computer: An historical perspective on the modern productivity paradox. *American Economic Review* 80: 355–61.

Greenwood, J., and M. Yorukoglu. 1996. "1974." Unpublished. University of Rochester.

Jovanovic, B., and R. Rob. 1990. Long waves and short waves: Growth through intensive and extensive search. *Econometrica* 58: 1391–1409.

Juhn C., Murphy K., and Pierce B. 1993. Wage inequality and the rise in returns to skill. *Journal of Political Economy* 101: 410–42.

Shleifer, A. 1986. Implementation cycles. *Journal of Political Economy* 94: 1163–90.

6   The Internet as a GPT: Factor Market Implications

Richard G. Harris

6.1   Introduction

The concept of a general purpose technology is subject to a variety of interpretations as discussed by Lipsey et al. in chapter 2. The original analytical models of Bresnahan and Trajtenberg (1994) and Helpman and Trajtenberg (chapters 3 and 4) emphasized the pervasive nature of the investment in complementary technologies that the introduction of a GPT induces, and the oft-cited examples of electricity and computers fit well with this intuition. Most economic historians, however, would also point to major developments in transport technology as leading to similar pervasive economic effects. The wheel, canals, clipper ships, steamships, and railroads all transformed economies and led to pervasive changes in the spatial organization of economic activity.[1] In particular, changes in transport technology led to increases in trade, and the associated gains from trade through specialization, greater scale economies, and the realization of comparative advantage. Much, if not most, of the growth in trade between spatially separated economic areas is a consequence of GPTs in transport technology. The theme of communications networks and their impact on international trade in goods was taken up in Harris (1995). In this chapter the focus is on communications networks interacting with international trade in services and the consequences for factor markets.

Innovations in communications technology such as the Internet—in particular, the ability to transmit voice, video, graphics, and large volumes of digitized data instantly, and at close to zero marginal cost, to large populations around the world—most certainly qualify as a GPT by almost

---

Comments of the members of the Canadian Institute for Advanced Research Growth and Economic Policy program and the referee are gratefully acknolwedged. All correspondence regarding the paper should be sent to the author at rharris@suf.ca.
1. Lipsey et al. discuss the historical literature on these and other forms of GPTs in chapter 2.

any definition. Here one aspect of that technology and its economic effects is discussed—the ability to remove the barriers to mobility of services. In the international trade literature a common distinction is drawn between services and goods. Commodities with a sufficient degree of durability and transportability are such that production can be divorced from consumption. Trade occurs via the transport of goods from the location of production to the location of consumption. Service transactions, on the other hand, are often characterized by the requirement that there be a double coincidence in both time and space of the proximity of the buyer and seller. The act of production cannot be distinguished from the act of consumption. Haircuts are a good example of a so-called pure service. However, many business services such as engineering and consulting services are not pure services but clearly come closer to the service end of the spectrum rather than the goods end. The major implication of the new communications networks (hereafter referred to generically by the term *Internet*) is that at reasonably low cost they break the necessity for buyer and seller to be at the same location, even though the coincidence in time may not be broken. Thus engineers located in one city can provide their services at any point on the globe given both the user and the engineer can communicate with each other via the Internet. In the language of the day the Internet creates a virtual meeting place, or in the language of international economics a form of "virtual mobility."

This is an idealistic view of this technology as it stands. Face-to-face communication still has considerable value, and in many services it may well be that transport technology (bringing parties physically together— e.g., baseball teams and fans) will continue to dominate electronic or virtual exchange in many situations. Nevertheless, it is evident much of what is referred to above is already occurring. The growth of the Internet has been nothing but staggering. A survey by Neilsen in July of 1996 report that as of end of 1995 there were 48 million users in the United States, 15 million in Europe, and approximately 78 million in the world as a whole.[2] The *Economist*[3] reports that the number of Internet hosts in the OECD has risen from less than 2 per thousand persons in 1991 to almost 16 per thousand by January of 1996. Business use is growing rapidly. In a survey by Neilsen business respondents listed as the top three uses of the Internet (1) to gather information, (2) to collaborate with others, and (3) to provide vendor support.[4] Clearly activities 2 and 3 correspond closely to

---

2. Reported at website http://www.euromktg.com/surostats.html
3. See the *Economist*, October 19, 1996, pp. 23–27.
4. See website http://www.nus.ir.choice/surveys

the notion of a service transaction either within the firm or external to the firm.

In this discussion of service transactions among unrelated buyers and sellers, it should also be emphasized that firm-specific networks, particularly within multinationals, are capable of serving a similar function within the firm. Different divisions of the firm can be located at a wide variety of locations and deliver services via and *Intranet* that have similar consequences to that of the Internet. In both cases service transactions between parties can be mediated over vast distances electronically both instantaneously and at low cost. These exchanges can be as long- or as short-lived as necessary, and they can involve the transmission of incredibly complex information such as that involved in accounting, engineering, R&D, or product service activities. What is the major economic consequence of this innovation? The emphasis in this chapter is that it will facilitate the enhancement of trade in business services across space. This will involve an increase in either intraregional trade or international trade. Another effect will be on factor markets, and in important ways on the market for skilled labor. Barriers to the mobility of skilled labor services, in part reflected the barriers to trade in business services that are intensive in the use of skilled labor services. If services become "virtually" mobile, (or alternatively become "traded"), this will result in shifts of the relative supplies and demands for skilled labor around the globe. This factor market effect may be among one of the most important consequences of the Internet.

Economists' traditional thinking about factor mobility[5] and factor returns can be summarized nicely in the following diagram familiar from the analysis of the specific factors model: Suppose that there exists a factor, called skilled labor, which might be allocated between two regions, a and b, if it were mobile. Initially, however, technology is such that it is impossible for this to happen. Each region produces a collection of goods with the resulting value of marginal product schedules for skilled labor depicted in figure 6.1 as VMP_a and VMP_b. The horizontal size of the box represents the world supply of the factor, the vertical axis the wages (value marginal products in each region), and point X the initial and exogenous endowment of these factors in each region. As drawn, region a has a higher wage for skilled labor than region b in the initial equilibrium. Now

---

5. The classic reference on the connection between factor mobility and trade is Mundell (1957) who recognized that trade and factor mobility are substitutes. In this chapter's model factor mobility and service trade are shown to be technologically induced by the same GPT, so they are treated as complementary to each other.

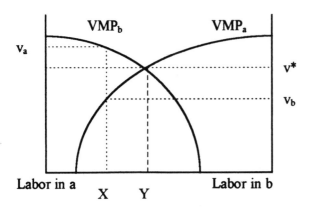

Figure 6.1
Wages and mobility consequences

imagine that a perfect and costless mobility machine is invented that allows skilled labor to sell its services in either *a* or *b*. The effect is that wages in both countries settle at v^* where (value) marginal products are equalized. Unfortunately, this model does not work very well at explaining why the skilled wage premium has risen in the industrialized countries (the initial high-wage countries), since it would predict that skilled labor services would be imported from the lower-wage countries, thus driving down the skilled wage in the high-wage country. For this reason this simple story about increased factor mobility and wages is not likely to fit the facts very well. By introducing a specific and costly form of factor mobility technology, in the form of a communications network, a story that fits the facts somewhat better is the purpose of this discussion.

There are at least three themes relating wages to introduction of a communications GPT that are worth exploring. First, there is the possibility that increased factor mobility may have played some role in the emergence of the skill premium in the 1980s and 1990s. The traditional explanation of skill-biased technological change has been challenged by those who seek to find explanations in the phenomenona of globalization.[6] Increased mobility of skilled labor achieved via a mobility-enhancing technology is an alternative explanation. Second, there is the possibility, emphasized by Helpman and Trajtenberg in chapter 3, following the "creative destruction" theme of Aghion and Howitt (1992), that a major shift in factor demands following the introduction of a GPT will reduce the demands for particular skill groups and increase that for others. Another

---

6. This debate is summarized nicely in the Symposium (1996).

way of saying this is that specific forms of human capital are obsolesced by the introduction of the new technology. In an international context some of those effects may be concentrated geographically. For example, if software engineers from India can deliver their services to the United States and Europe via the Internet, then this might erode the high wages of software engineers in these countries while at the same time promoting the expansion of their software industries. The third theme is that of agglomeration economies. In much of the recent trade and geography literature, following the initial contribution of Krugman (1991), trade between spatially separated communities gives rises to a trade-off between scale economies and increased transport costs. Unlike the trade and geography literature, what is possible in this instance is that low communication costs facilitates the mobility of skilled labor services while not necessarily compromising on better scale economies. This in turn can lead to agglomeration of economic activity via the concentration of the use of factor services at a particular location. A common theme in the popular literature on communications technologies is the question whether due to these effects small countries become winners or losers from this technology. If delivery of skilled labor services were withdrawn from the local economy, and their delivery shifts to the large countries, it may indeed be that the net effect would be detrimental to the national welfare of the small country.

Both sets of issues will be explored in a simple general equilibrium model of trade, emphasizing the impact of a mobility-enhancing GPT. To keep matters reasonably simple, a comparative static form of analysis will be used comparing the pre-Internet situation with the post-Internet situation. Another major simplification will be the restriction of the analysis to the integration effects of the Internet on economies, or regions, facing exogenous terms of trade set in a third country (rest-of-world), a technique commonly used in the customs union literature. This allows considerable simplification and attention to the factor market consequences, although terms-of-trade effects are obviously missing from the analysis.

6.2 A Model of Trade among Regions

The first model is of an open economy consisting of multiple geographically distinct regions (or cities). Each region produces three goods and uses two factors, skilled and unskilled labor. Two of the goods are traded, and one—services—is not traded. Both skilled and unskilled labor, denoted S and L, are assumed to be physically immobile across regions.

There are two traded goods produced in each region. One good will be referred to as the M good (indexed as sector 1), or manufacturing sector, and production of this good uses both skilled and unskilled labor. Unskilled labor is a factor input specific to manufacturing—thus we can think of this sector as a catchall for traditional traded goods industries. The M sector is a competitive constant returns industry using unskilled and skilled labor, with a production function

$$Y_1 = F(L, S_1). \tag{1}$$

There is a second industry, sector 2, which will be referred to as the T sector, or technology sector. The T sector has also competitive constant returns but uses only business services as inputs.[7] Given an n-vector z of business service inputs the output Y_2 of the T sector is

$$Y_2 = \left(\sum_{i=1}^{n} z_i^\rho \right)^{1/\rho}. \tag{2}$$

The third nontraded sector is business services, or the B sector, which is monopolistically competitive, producing differentiated business services that use skilled labor as the only input. The production function for service good i is given by

$$z_i = \begin{cases} \dfrac{1}{b} s_i & \text{if } f > 0, \\ 0 & \text{otherwise,} \end{cases} \tag{3}$$

where s_i is the variable input of skilled labor to service good i and f is a fixed input of skilled labor, which is independent of the scale of output. Note that skilled labor is used directly in sector 1 but only indirectly in sector 2, since it is used to produce an intermediate composite input, business services, which are the sole inputs to that sector. Within a region, skilled labor is mobile between the M sector and the B sector.

Consider a single location with an endowment of unskilled labor L and skilled labor S. Both M and T goods are sold at fixed world prices p_1 and p_2. The price of service input i is q_i. The symmetric MCE equilibrium has n business services produced in quantities $z_i = z$ and with prices $q_i = q$. As price equals unit cost in sector 2:

$$p_2 = qn^\lambda \tag{4}$$

---

7. This specification follows that of Ethier (1979).

Table 6.1
Summary of notation

p_1	Price of good 1—M sector
p_2	Price of good 2—T sector
v	Price of skilled labor
w	Price of unskilled labor
q	Price of representative service good
n	Number of differentiated service inputs to manufacturing sector
S_1	Skilled labor used in M sector
S_2	Skilled labor used in B sector
S	Aggregate endowment of skilled labor
L	Aggregate endowment of unskilled labor specific to M sector

with

$$\lambda = \frac{\rho - 1}{\rho} < 0 \quad \text{and with restriction that} \quad \rho < 1. \tag{5}$$

The wage paid skilled labor in services will be referred to as v_2; sector 2 in the factor market sense can be usefully thought of as an amalgamation of the nontraded service sector and the traded final good T. The price markup rule in the representative service firm is

$$q = \frac{1}{\rho} b v_2, \tag{6}$$

since skilled labor is the only variable input to services. Substituting in (4), we have

$$p_2 = \frac{1}{\rho} b v_2 n^\lambda. \tag{7}$$

Solving for service sector wages as a function of the price of T and the degree of input differentiation gives

$$v_2 = \frac{\rho p_2}{b n^\lambda}. \tag{8}$$

Taking logs of (8) yields

$$\log v_2 = \log\left(\frac{\rho p_2}{b}\right) - \lambda \log n. \tag{9}$$

Equation (9) gives the value of the average product of skilled labor in the service/technology sector, and this is increasing in the level of input differentiation. Since it is also true that in zero profit monopolistically competitive equilibrium the number of service varieties, n, adjusts such that price equals average cost on each variety, then

$$q = bv_2 + \frac{v_2 f}{z}. \tag{10}$$

Using the markup rule and price equal average cost, we solve for equilibrium scale z in the representative service sector as

$$z = \frac{v_2 f}{bv_2} \frac{p}{1-p}$$

$$= \frac{f}{b} \frac{p}{1-p}. \tag{11}$$

Total skilled labor use in sector 2 is given by

$$S_2 = nf + nbz. \tag{12}$$

Solving for the number of service varieties as a function of S_2 gives

$$n = \frac{S_2}{f + bz}$$

$$= \frac{S_2}{f}(1-p). \tag{13}$$

The number of varieties is linear in the supply of skilled labor to the service sectors.[8] Substituting for n in the service sector wage equation, we have

$$\log v_2 = k - \lambda \log S_2 + \lambda \log f, \tag{14}$$

where k is a constant equal to $\log(pp_2/b) - \lambda \log(1-p)$. L is a specific factor in sector 1, so we also have the wage equal marginal products condition (or price equal unit cost) in that sector:

$$\begin{aligned} v &= p_1 F_S(L, S_1), \\ w &= p_1 F_L(L, S_1) \end{aligned}. \tag{15}$$

---

8. Note that n must be greater than 1 which places a lower bound on S_2. In the diagrams this problem is not apparent because of their scale, so the productivity curves for sector 2 are zero for $n = 0$. If f were very large, productivity in services would only be positive for some positive finite value of S_2 sufficient to cover more than the fixed cost of setting up at least one firm.

Factor market clearing requires that

$$S_1 + S_2 = S. \tag{16}$$

Factor market clearing in a single market is depicted in figure 6.2 (the traditional specific factors diagram) with the horizontal axis representing the total available supply of skilled labor in a single region. On the left side a downward-sloping value marginal product of skilled labor schedule (VMP_1) is drawn and on the right, a rising value of average product (AP_2) schedule for skilled labor in the technology/service sector. The equilibrium allocation of skilled labor is determined by intersection of the two schedules.

With increasing returns to scale and monopolistic competition in services, an increase in the quantity of skilled labor allocated to services, S_2, raises the average product of skilled labor in sector 2 in a diversified equilibrium in which both M and T goods are produced. The skilled wage increase is a result of the Smithian effect of increasing specialization (larger n) in the service sector. An increase in the total supply of skilled labor will raise the price of skilled labor v and reduce the allocation of skilled labor to the manufacturing sector provided that the economy produces both M and T. This would necessarily also reduce the wage of unskilled labor, w. This argument follows as the demand for skilled labor in the economy as a whole, in the long run, is increasing in the skilled wage v.[9] Holding the stock of unskilled labor constant, a region with a larger supply of skilled labor will have a higher degree of specialization in services, a higher-skilled wage, and a lower-unskilled wage. Holding the size of the skilled labor forces equal, the market with the larger unskilled labor force will have a lower-skilled wage and a higher-unskilled wage (assuming a nonspecialized equilibrium). The model yields different predictions on the relationship between the factor price ratio v/w and factor supply ratio S/L than the standard competitive model. In a diversified

---

9. In figure 6.2 the VMP curve in sector 1 cuts the AP curve in sector 2 from above. There are two explanations for this configuration. First, it reflects the fact that the economy is diversifed in both M and T goods. If the AP_2 curve were to cut the VMP curve from below, the economy would specialize in the production of T. Second, the interior equilibrium allocation has a type of Marshallian stability familiar from the specific factors model. In the short run, if labor supplies are fixed in the two sectors at an allocation other than the intersection of the two schedules, then wages will differ in the short run between the two sectors. As drawn in figure 6.2, to the left of the intersection point wages in sector 1 would be above those of sector 2. This would lead over a longer run for labor to move from sector 2 to sector 1 and for wages to equalize. A similar argument is valid if the short-run allocation is to the right of the intersection point.

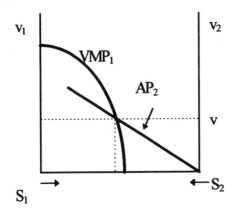

Figure 6.2
Equilibrium in the market for skilled labor

equilibrium v/w is positively related to S/L. Thus regions with higher proportions of skilled to unskilled labor will have a larger skill premium.

A *nation* or *regional economic bloc* might be thought of a as a collection of regions among which *firms* are potentially mobile in the time frame of interest.[10] In this model it is assumed that all firms earn long-run zero profits. In the initial situation service firms, or more accurately the services they provide, are region specific. Thus due to the nature of their product, they both sell their outputs and purchase their inputs in the same regional market. With m regions all with the identical endowments of (L, S) of each type of labor, the equilibrium will be one in which wages w and v are equal across all regions. We can think of the economy as a whole as being characterized by the triple (m, L, S).[11]

---

10. This is slightly different than the usual definition in which a country is defined as a collection of regions among which factors are mobile. The definition used here conforms more closely to that used in the regional economics literature in which firms (and thus capital) are mobile across regions but the labor force is not.

11. In this model it is useful to think of *fixed cost density* as given by S/f and L/f—the ratio of the size of the labor force to the fixed cost per firm. If this ratio is large, we say density is high—likewise if it is small, density is low because either S is small or f is large. Since $F_S(L, S_1)$ is homogeneous of degree zero, and n is linear in S_2/f, the model can be solved in the variables L/f, S_1/f, and S_2/f. An increase in fixed cost density will increase the horizontal size of the box but leave the two productivity schedules unchanged, and thus raise equilibrium wages and increase the size of the service sector. Consequently we can say that *more dense regions* will have higher wages to skilled labor. This can be either because they have a larger population of skilled workers or because geography is such that fixed costs per firm are lower. The benefits to increasing returns to scale show up either way. An economy of given resources but with more regions will have lower density per region and thus in equilibrium will have lower wages.

6.3 Introducing the Internet

The Internet (or Net for brevity) is a communications network that covers all regional markets. The introduction of the Net allows any service provider to sell into any of the m regional markets. Service activities are thought of as being uniquely intense in their communications requirements due to the necessity of bringing buyer and seller together. The presumption is that virtual meetings via the Internet are a perfect substitute for face-to-face meetings. Prior to the Internet we can think of the service and technology sectors as existing in a state of *communications autarky*—communication is possible within regions but not between regions, a region being defined as a geographic area sufficiently small that service activities via face-to-face meeting technology and without the Internet. For similar reasons the Net also allows any service firm to purchase the services of skilled labor from any other regional market. Thus the Net effectively integrates both the regions' business services market and the skilled labor market. From a mobility point of view the Net has created *virtual mobility* of both services and skilled labor.

Prior to the Net the fixed costs of setup per firm within a single market are vf. From a firm's perspective the costs of using the Net are all fixed costs. To be on the Internet, there is an additional fixed cost to each service firm of an amount $v\psi(\cdot) - vf$, where the fixed cost per user firm is a function which depends potentially on a number of factors, including number of users, number of regions, geography, type of technology used, market structure, and so forth. We will start with the simplifying assumption that ψ is constant and independent of all these factors. It is also assumed that the Net is produced using inputs only of skilled labor.[12] Moreover the factor demands created as a consequence of the production of the Net are proportional to the number of user firms in each region's market. Note that as a consequence of these assumptions skilled labor gains not only as a user of the Net but also as the sole factor input to production of the Net. The total resource cost of providing the Net across the economy is linear in the number of users. At one level these assumptions imply (1) that there are aggregate constant returns to scale in producing the Net and (2) that the pricing of Net services is done on an

---

12. This factor intensity assumption is obviously unrealistic in that is excludes capital costs. It is a useful approximation in a two-factor model of skilled and unskilled labor. It is fairly clear that ICT technology is complementary with increased skill levels.

average cost basis. Neither of these assumptions is particularly realistic, but they are a useful starting point.[13]

6.3.1 Factor and Real Income with the Internet

The impact of the Net is to allow service firms to sell their services to any T sector firm, no matter where located. This effectively integrates economically the T and B sectors nationally. Since both are users (direct and indirect) of only skilled labor, the actual location of T sector output across regions is indeterminate. Given the symmetric CES input structure, sector T purchases services from all regions business service firms. Thus, in aggregate, the number of differentiated service inputs rises from n to mn on impact. At the same time fixed costs per service firm go from f to ψ. The new equilibrium in a representative region is characterized by the equations

$$n = \frac{mS_2}{\psi + bz}$$

$$= \frac{mS_2}{\psi}(1 - \rho), \tag{17}$$

$$\log v_2 = k - \lambda \log S_2 - \lambda \log m + \lambda \log \psi, \tag{18}$$

$$v_1 = p_1 F_S(L, S_1), \tag{19}$$

$$S_1 + S_2 = S, \quad \text{and} \quad v_1 = v_2. \tag{20}$$

The net impact is represented in figure 6.3, which compares the pre-Internet (communications autarky) equilibrium, point A, and post-Internet equilibrium outcome, point B. The critical ratio to compare is f versus ψ/m, which determines whether the AP locus shifts up or down in response to the introduction of the Net. If ψ/m is smaller than f, then the value average product curve in sector 2 shifts up. This raises the skilled wage and increases the resources devoted to the service sector, and thus the size of the high-technology sector. At the same time the wage of unskilled labor is reduced as skilled labor is pulled out of the traditional goods sector and into the service sector as a consequence of the virtual mobility effects of the Internet. The model predicts an increase in wages

---

13. It is perhaps useful to emphasize the distinction here between transport costs and communication costs. The former is a marginal cost which is positive, and thus total transport costs are increasing in the volume of trade. The marginal costs of communications, on the other hand, is assumed to be zero—the costs of a communications network are all fixed costs.

The Internet as a GPT

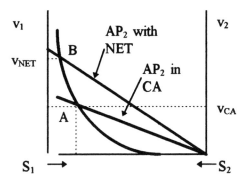

Figure 6.3
Wages and the introduction of the Internet

to skilled labor, a decrease in unskilled wages, and an increase in the skill premium as a result of this particular form of technological change.[14] Note that the technological change is not costless. Firms must pay $v(\psi - f)$ in order to be on the Net. In the new equilibrium the region as a whole is using $n(\psi - f)$ units of skilled labor as inputs to the production of Net services.

Post-Internet there is now trade between regions of business services where previously there was none. The volume of service imports in a single region is given by $S_2(m - 1)/m$ and in the economy as whole business service trade embodies $(m - 1)S_2$ of skilled labor inputs. The productivity benefits of communications are the gains from trade that accrue from increased specialization in the provision of business services. These productivity gains are offset against the cost of the Internet. The net effect is to raise wages to those benefiting from the innovation in virtual mobility technology, and reducing wages to those that are immobile. The

---

14. It is useful to contrast this explanation of the skill premium with two more conventional explanations. The first is a factor-biased shift in production functions against unskilled labor, which raises the return to skilled labor at any exogenous fixed supply of skilled and unskilled labor in the economy. In this model the technological change is solely in the communications technology or "factor mobility" technology. The increase in mobility unambiguously increases the wages of skilled labor due to the productivity improvements in the T sector captured by skilled labor. A corollary of the model is that in the unskilled intensive sector (the M sector) the ratio of skilled to unskilled labor falls as the skill-intensive sectors expand in the economy as whole.

The second alternative explanation comes from traditional 2×2 trade theory and the Stolper-Samuelson theorem. In that model the wage premium to skilled labor rises if technological change is sectorally biased in favor of the skill-intensive sector and terms of trade are exogenous. In the model developed here there is also asymmetric technological change. The productivity improvements due to the GPT occur entirely in T sector which is intensive in the use of skilled labor.

increase in wage inequality in thus correlated with an increase in trade, although both are caused by the introduction of the GPT.

The effect on GDP of each region is unambiguously positive provided that the skilled wage increases. The regional GDP is equal to factor income given by $wL + vS$. The total returns to unskilled labor is the area under the VMP_1 curve less the area wS_1, the wage bill of skilled workers. The losses to unskilled labor from the introduction of the GPT are more than offset by the increase in returns to skilled labor. The introduction of the Net, provided that it raises skilled wages, therefore raises aggregate real income and productivity.

6.3.2 Economies of Size in Internet Provision

There are good reasons of course to believe that the cost per user of providing the infrastructure of a communications network are rapidly decreasing in the number of users. This is an additional form of scale or size economy beyond firm-level scale economies, which is important at the aggregate level, both intraregionally and interregionally. The fixed cost f represents the fixed cost of setting up a firm. The function ψ represents f plus network costs per user. Assume that we can proxy the number of users, u, by the total number of service firms on the Net; then $u = nm$. If ψ is a decreasing function of u, the situation is such that for small u the fixed costs per firm are very high and thus average productivity in sector 2 is very low. As the number of firms connected to the Net increases the fixed costs per user diminish rapidly.

This gives rise to the configuration depicted in figure 6.4 in which the AP_2 curve in the presence of virtual service market integration is well below the communications autarky productivity curve for low values of S_2, and then rises rapidly as costs per user fall. For sufficiently large S_2 the new services productivity curve lies above the same curve under communications autarky. This type of model gives rise naturally to multiple equilibrium. The communications autarky equilibrium is denoted by point A. In the presence of the Net there are two equilibrium represented by points B and C in figure 6.4. Point B represents the low-level equilibrium and gives rise to lower wages and a smaller service sector than does communications autarky. Point C is the high-level equilibrium and represents a point with wide-scale adoption of the technology and high wages to skilled workers.[15] Economic theory is not good at predicating how the

---

15. The third intersection point of AP_2 with NET and VMP is unstable, and therefore it is assumed not to be of interest in the long run. See note 9 above.

The Internet as a GPT

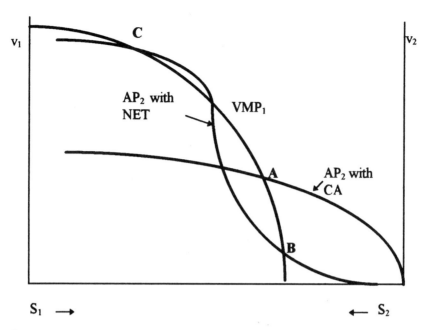

Figure 6.4
Communications autarchy versus virtual integration

shift from one equilibrium to another occurs, but in this case the adoption of the GPT brings a large change in wages and resource allocation across sectors; it also creates losers among the specific factor owners in the traditional goods sector who do not benefit from the additional mobility.

6.4 Regional Differences

In this section differences in region size are examined. A larger region, as measured by a higher ratio of S to f, will have a higher-skilled wage and lower-unskilled wage than the smaller region. What would be the impact of integrating two regions (one large and one small) with a communications network? To investigate this issue, consider first the case of costless integration, and then examine what happens when the costs of running the communications network are introduced.

Suppose that skilled labor services are perfectly mobile across the two regions but that sector M specific unskilled labor is not. In this world v is the same in both regions and q is the same in both regions. Since v is the same and all M sector firms in both regions sell their output at the same price p_1, it must also be the case that w is the same in both regions and consequently there is factor price equalization (FPE). Prior to allowing trade in skilled labor services, the initial equilibrium is one of differences across the two regions in skilled and unskilled wage rates. Because of increasing returns to average productivity in the services sector, there is a strong presumption that integration of the services market might raise skilled wages in both regions. However, there are two forces present, and they will not always work in the same direction:

1. Equalization of marginal products holding the pattern of specialization constant (the factor efficiency effect).
2. Reallocation of industries across regions (the specialization effect).

The first effect is identified by the requirement that the increasing returns to scale industry (IRS) industry operate in both regions but skilled wages are equalized by the movement of skilled labor.[16] Equality of skilled wages can only occur with identical productivity in both regions, and this can only occur if the size of the technology/services sector is the same in both regions. For skilled workers small countries gain and large countries lose. The specialization effect allows for elimination, or more accurately consolidation, of one of the two service sectors thus eliminating duplication in the provision of identical varieties, and raising productivity in both countries. The net impact on wages depends on the relative importance of these two factors.

6.4.1 Factor Efficiency Effect

Factor efficiency raises skilled wages in the small country and lowers skilled wages in the large country relative to the initial situations, since skilled wages are initially lower in the small country. Returns to the unskilled specific factor in the small country fall. The technology/services sector expands in the small country and contracts in the large country, while returns to unskilled labor in the large country rise. The large country, given its relative abundance of the specific factor in sector 1, and thus comparative advantage in that sector, attracts the mobile factor to the manufacturing sector which raises the sector-specific rents (unskilled wages) and lowers the wage of skilled labor.

---

16. This allocation is determined analytically by duplicating the technology/service sector in both regions.

6.4.2 Specialization Effect

Suppose that wages have been equalized by the process of equalizing average productivity in the service sector across both regions. At these wages and factor allocation, services can now be consolidated into one sector. This is equivalent to doubling the amount of skilled labor available to that sector. To restore equilibrium, wages of the mobile factor rise in sector 1 (in both regions), and the new sector 2 is larger than it was prior to integration and with higher productivity. Unskilled wages (specific factor rents in sector 1) fall in both regions.

Enhancing mobility in and of itself without changing the pattern of specialization tends to equalize wages, in this case lowering them in the large country and raising them in the small country. The specialization effect of mobility tends to raise skilled wages. Consequently a zero-cost Internet will have an ambiguous effect on wages in the case of asymmetric region sizes. Adding the cost of supplying the Net complicates this picture in the predictable direction. Additional fixed costs per service firm raise the average equilibrium size of a service sector firm, reduce input differentiation, and thus average productivity in technology/services. On balance, the effect of additional input variety on T sector productivity, which comes from integrating the two service markets into a single market, may easily overcome the additional resource cost of providing the Net raising skilled wages in both regions.

6.4.3 Agglomeration and Small Regions—Can They Be Made Worse Off?

In our model, in the absence of the Net there are agglomeration effects in large regions due to scale economies in services. Suppose that skilled labor can migrate from low-wage (small) to high-wage (large) regions. This migration will continue to occur until a sufficient amount of skilled labor has left the small region and its M sector can support a wage to skilled labor equal to that in the larger region; however, wages to unskilled labor will fall due to the out-migration of skilled labor. Now introduce the Internet. The Internet can eliminate the scale disadvantage of small regions in producing services if the small region is attractive for other reasons, such as environment or the nature of local public goods. The Internet then can potentially lead to in-migration of skilled labor to the region.

In regions divided by international borders, migration is likely to be controlled. A small country in communications autarky simply suffers a lower-skilled wage than a large country. The introduction of the Net raises wages to the mobile factor and reduces that of unskilled labor.

If there are scale economies in the provision of Net infrastructure, then there economies are potentially interregional in scope. A small country joining a large country network thus benefits from the scale economies achieved by the large region. This is yet another reason small regions are not disadvantaged relative to large regions by the introduction of a mobility-enhancing GPT.

6.5 Internet Adoption

The question of what circumstances may give rise to the introduction of the Internet, given the technological knowledge necessary to construct one, is a very complex issue. Complicating the analysis is the fact that a communications network is associated with a number of the features commonly associated with public goods. There is further the distinction that can be drawn between the provision of network infrastructure and individual decisions to invest in accessing the network.

In our model these decisions are collapsed within the representative firm framework to one in which all firms are either on or off the network. Rather than analyze in detail the incentives that impinge on that collective adoption decision to introduce a communications network, a much simpler approach is taken by asking what individual incentives exist to use a network, assuming that it can be provided. Consider the hypothetical question of whether a firm, given the option to purchase access to the Net at a price $v\psi$ would choose to do so when all other firms are currently not on the Net. If all firms give an affirmative answer to this question it is more likely that adoption would occur. As such this is a weak sufficiency test for introduction of the Net as a GPT.

A firm that chooses not to go on the Net sells only into the local market. Consider the case of symmetrically sized regions with equal factor prices. Given marginal cost is the same for a firm with or without access to the Net, the decision to purchase access to the Net has no effect on the markup on cost in setting price, and thus profits per unit sales are the same with or without the Net. What does change, however, it market size. Let z^* denote equilibrium sales of the representative service firm in communications autarky prior to the introduction of the Net. There are given by equation (11) and equal to $(f/b)(\rho/(1-\rho))$. Note that the initial level of sales *per market* is set by the level of initial fixed costs. Going on the Net yields expected sales at a level of mz^*. Then the expected profits from going on the Net are given by

$$\Pi^{new} = \frac{1-\rho}{\rho} bvmz^* - \psi v \qquad (21)$$
$$= v\{mf - \psi\}.$$

Consequently, when $f > \psi/m$, the adoption of the Net will presumably occur. In this case, comparing fixed costs suffices because initial fixed costs in equilibrium proxy for expected variable profits.

There are three distinct channels of influences on the adoption of the new GPT such as are embodied in the three parameters f, m, and ψ and the wage v:

1. Larger gains in market size from Net access (larger m).
2. Reductions in the cost of accessing the GPT (lower ψ).
3. Higher wages in the service industries (higher v), which imply higher variable profit margins, given that $mf > \psi$.

In this model the conditions for skilled wages to rise with the introduction of the GPT are also coincident with those for the expected private profitability of the individual firm's adoption decision. Since firms in the long run earn zero profits, rents earned on the GPT are necessarily transitory. In the long run all the gains are captured by skilled labor in the form of higher wages. The losers are unskilled labor in the symmetric case.

This framework can be used to generate temporary productivity declines in a dynamic model, as emphasized in the Helpman and Trajtenberg GPT models. Productivity declines occur in transitions when resources are devoted to developing complementary technologies to the GPT. In the Internet case the removal of a communications autarky among service firms and their customers is likely to proceed in stages. At the first stage when the Net is introduced, service firms will incur the fixed cost ψ. The benefits in terms of effective use of the increased market size, that is, higher m, take time to materialize. One can think of it as a sequence of temporary equilibria, all with the same Internet cost function, $\psi(\cdot)$, but indexed by *effective* market penetration, or m^*, which rises slowly to full integration, or $m^* = m$. The average product curve in services for each m^* then becomes.

$$\log v_2 = k - \lambda \log S_2 - \lambda \log m^* + \lambda \log \psi. \qquad (22)$$

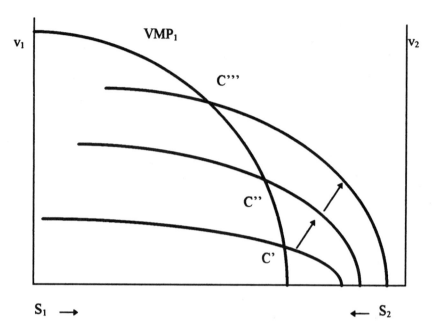

Figure 6.5
Internet adoption and temporary productivity declines

The situation is depicted in figure 6.5 by the shift from the CA equilibrium of figure 6.4 to C', then to C'', and C''' as m^* rises toward m with increased use. The initial post-Internet equilibrium can be below the pre-Internet equilibrium point. (compare points A and B in figure 6.4). The dynamics therefore go from an initial drop in skilled labor's wages and productivity to a long rise in productivity over time.[17]

6.6 Conclusion

This chapter has constructed a model of interregional trade affected by a GPT in the form communications network, or Internet. Such a communications network can be used to facilitate trade among regions in business services and the factor services of skilled labor. Setting up the network is costly, however, and the technology of communication is such that all

---

17. Note that drawn, the AP_2 curves become positive at increasingly smaller values of S_2. This dynamic reflects the fact that fixed costs per firm are very large, and that there is minimum effective size of the service sector at which productivity is positive. However, that size becomes smaller as the fixed costs are spread over a large portion of the market, namely as the ratio m^*/ψ rises.

costs of communication are fixed costs per firm. On the factor market side the model assumes that there are two factors: skilled and unskilled labor immobile across regions; business services are provided only by skilled labor. By this assumption, skilled labor is a complementary input to the mobility, enhancing GPT. The main conclusions are as follows:

1. The introduction of a communications network that facilitates mobility of business services also enhances the "virtual" mobility of skilled labor. As a result the wage premium to skilled labor increases and so does the absolute real wage of skill labor.

2. While small countries may have lower absolute labor productivity in technology and services prior to the introduction of the communications GPT, after the Net is in place their productivity levels equal to those of large countries.

3. Globalization is manifested in increased volumes of interregional trade in services and in a larger service sector.

The chapter provides a straightforward theoretical framework that includes reduced communications costs as a common technological factor that drives the globalization of a large service sector resulting in increased productivity and increased income inequality. The GPT paradigm appears to be a powerful organizing framework in which to think about these unique properties of the new communications technologies.

References

Aghion, P., and P. Howitt. 1992 A model of growth through creative destruction. *Econometrica* 60: 323–51.

Bresnahan, T., and M. Trajtenberg. 1995. General purpose technologies: Engines of growth? *Journal of Econometrics* 65: 83–108.

Ethier, W. 1979. Internationally decreasing costs and world trade. *Journal of International Economics* 9: 1–24.

Globerman, S. 1996. The information highway and the economy. In P. Howitt, ed., *The Implicatioins of Knowledge-Based Growth for Micro Economic Policies*. Calgary, Canada: University of Calgary Press.

Grossman, G., and E. Helpman. 1991. *Innovation and Growth in the Global Economy*. Cambridge: MIT Press.

Harris, R. 1995. Communications costs and trade. *Canadian Journal of Economics*, special issue, S46–S75.

Krugman, P. 1991. Increasing returns and economic geography. *Journal of Political Economy* 99: 483–99.

Markusen, J. 1989. Trade in producer services and other specialized intermediate inputs. *American Economic Review* 79: 85–95.

Mundell, R. A. 1957. Ínternational trade and factor mobility. *American Economic Review* 47: 321–35.

Symposium 1995. Income Inequality and Trade. *Journal of Economic Perspective* 9: 15–80.

7 Chemical Engineering as a General Purpose Technology

Nathan Rosenberg

7.1 Introduction

This chapter examines the discipline of chemical engineering as a general purpose technology and looks at the relationship between this particular GPT and the rise of its application sectors—primarily petroleum and petrochemicals. In this context the chapter explores the notion that it is useful to conceive of a set of ideas as constituting a GPT.[1]

But the chapter is also motivated by another interest. There is widespread agreement that technological change is, at bottom, some kind of learning process. The big question is exactly how that learning takes place—specifically, how does society acquire the new knowledge and capabilities that constitute technological change? Economic theorists and economic historians would both agree that growth of knowledge needs to be examined, at least as a starting point, as the outcome of purposive investments by economic agents.

The author wishes to express his gratitude to Moses Abramovitz, Ashish Arora, Tim Bresnahan, Alfonso Gambardella, Ralph Landau, Christophe Lecuyer, and Scott Stern for many helpful, and occasionally invaluable, suggestions and criticisms. He has also been the considerable beneficiary of comments from the Science, Technology, and Economic Growth Workshop at Stanford University and the Program in Economic Growth and Policy of the Canadian Institute for Advanced Research.

1. For a discussion of general purpose technologies, see T. F. Bresnahan and M. Trajtenberg, General purpose technologies: "Engines of growth?" in *Journal of Econometrics*, 1995, pp. 83–108: "The central notion is that, at any point of time, there are a handful of 'general purpose technologies' (GPTs) characterized by the potential for pervasive use in a wide range of sectors and by their technological dynamism. As a GPT evolves and advances it spreads throughout the economy, bringing about and fostering generalized productivity gains.... This phenomenon involves what we call 'innovational complementarities' (IC), that is, the productivity of R&D in a downstream sector increases as a consequence of innovation in the GPT technology. These complementarities magnify the effects of innovation in the GPT, and help propagate them throughout the economy." (p. 84)

In the new growth theory[2] the entire learning process, and the conditions on which success and development depend, are represented by inserting two variables into its equations: investment in R&D and investment in tangible capital (together with some observations on increasing returns and learning by doing). The actual progress of technology remains an unopened black box, but the central point is that a new blueprint today spills over to lower the cost of future blueprints. There is, in other words, an intertemporal externality.

The story that will be told here goes a certain distance toward analyzing the contents of that black box in the context of chemical engineering. No claim is made that the entire process of learning—of technical progress—has been rendered endogenous, but it is hoped that the questions of explanation have been posed at a deeper level. It is argued further that, even when one accepts the abstractions of the new growth theory and the centrality of the variables on which it focuses, effective policy formulation still requires knowing how to implement that abstract knowledge under practical conditions. Alternatively put, without the presence of the conditions examined here, the actual effects of the variables identified in the new growth theory remain somewhat uncertain. Perhaps more fully endogenizing the growth of knowledge, on which technological change depends, calls for a joint enterprise involving theorists, historians, and engineers.

According to the new growth theory all inventive activity reduces the cost of future inventive activity in the same way. What the evidence introduced here really seems to suggest, however, is that a particular type of innovation—in this case the creation of a new disciplinary framework—determines the parameters of the degree of intertemporal spillover. Without the formulation of the idea of unit operations, which is so central to chemical engineering, a tremendous amount of technological knowledge could not have spilled over into future technology. That knowledge would simply have been used in the creation of an idiosyncratic and singular product or process. However, with the development of the concept of unit operations and the codification of such knowledge in textbooks, a given amount of inventive effort led to a potentially larger spillover to future inventors. Thus, the concept of unit operations created an intertemporal spillover that reduced the cost of future inventive activity. In this sense there is a strong mutual influence between the endogeneity of the new growth theory and the phenomenon of general purpose technologies.

---

2. See in particular the seminal paper by Paul Romer, Endogenous technological change, *Journal of Political Economy*, 1990.

This is the larger context of the present focus on the growth of one particular engineering discipline, chemical engineering. The reasoning is straightforward. If one is interested in how technological learning accumulates, one needs to look carefully at the engineering professions. Most engineering professions—such as chemical, electrical, and aeronautical engineering and metallurgy—are relatively new disciplines, none of them much more than a hundred years old. These disciplines came into existence in response to the emergence of new industries, going back to the so-called second industrial revolution of the late nineteenth century. The engineering discipline now called "computer science" is of course of much more recent vintage.

The engineering disciplines need to be thought of as the repositories of technological knowledge, and their practitioners as the primary agents of technological change in their respective industries. The growth of useful new technological knowledge is largely the product of what goes on inside these engineering disciplines. In the chemical sector the chemist may be the main agent in the development of new products, but the design of the manufacturing technology for new products, as well as improved technologies for manufacturing old products, is in the hands of the chemical engineers.

It turns out not to be as easy as had originally been anticipated to explain observed behavior and outcomes in terms of the deliberate commitment of resources, on the part of maximizing agents, to capture available rents. The story of chemical engineering as a GPT is full of unintended and unexpected benefits. There was plenty of rent seeking, especially in the behavior of oil companies that moved downstream into the chemical industry when that industry was transformed into a petrochemical industry. There was also a good deal of proprietary behavior, such as patenting, although ironically the most important patent—Houdry's patent on catalytic cracking which turned out to be a major technological breakthrough—was instigated by the reluctance of oil companies to pay high patent royalty fees.

Indeed it is difficult not to come away amazed by the petroleum industry's failure to have pushed proprietary interests any further. How does one account for that failure? There are a number of possible explanations:

1. Universities played a prominent role in allowing useful new knowledge to become quickly public.

2. Much of the crucial information on chemical process design was not patentable.

3. Petroleum firms are big and, largely, price insensitive. The emerging GPT under consideration was an increasing returns technology. Demand was growing so rapidly that firms could be insensitive to price. The "name of the game," clearly was to increase output in response to a rapid growth in expected demand throughout the interwar years. (The behavior of firms in defense procurement may provide a useful analogy.)

This perspective of chemical engineering as a general purpose technology may be compared with GPTs that have been identified as specific forms of hardware: steam engines, machine tools, dynamos, computers, and so on. It suggests that the concept of a general purpose technology should not be confined to hardware. Indeed, a discipline that provides the concepts and the methodologies to generate new or improved technologies over a wide range of downstream economic activity may be thought of as an even purer, or higher-order, of GPT.

This should not be controversial. A steam engine or a dynamo is not a technology; they are examples of tangible capital equipment. "Steam" or "electricity," meaning bodies of knowledge about how to produce steam or electricity and to use them as sources of power or light, in steam engines or dynamos, are technologies. Similarly chemical engineering is a body of knowledge about the design of certain technologies. More specifically chemical engineering is a body of knowledge about the design of process plants to produce chemical or other products whose production involves chemical transformations. Since chemical engineering provides essential guidance to the design of a very wide range of plants, it may usefully be thought of as a GPT. Furthermore there has been both a vertical and horizontal dimension to the externalities that were generated. The emergence of chemical engineering meant that downstream sectors experienced lower invention costs. But, in addition, there was a powerful horizontal externality, in the sense that the vast market for petroleum has shaped the development of petrochemicals through the intermediation of chemical engineering.

7.2 Relationship of Chemical Engineering to the Science of Chemistry

The first step in this exploration must, necessarily, be a deck-clearing operation. In order to understand the role of chemical engineering in technological change, it is necessary to deal with a widely held view that chemical engineering, like other engineering professions, is simply applied science—in this case applied chemistry. It is suggested that chemical engineering can most usefully be thought of as a body of technological knowledge that is not reducible to applied science, nor did it have its historical origins in science, although at a later stage in its development it began to draw upon certain realms of science and, in that sense, became more scientific. Chemical engineering is, rather, a body of useful knowledge concerning certain manufacturing processes that can not be derived by deduction from or by simple extension from the established sciences.

Although the science of chemistry became a major source of new chemical products, beginning with Perkin's immensely important opening up of the field of organic chemistry in the second half of the nineteenth century, its series of breakthroughs provided very little guidance to the process technologies that would be essential to the manufacture of chemical products in the twentieth century. Chemical engineering emerged early in the twentieth century as a separate body of knowledge that could guide the design as well as the operation of chemical process plants, including plants producing well-established products, such as ammonia, in addition to new ones. The rapid expansion of chemical engineering in the twentieth century, however, was not so much due to the late-nineteenth-century growth of the synthetic dye industries but to other industries that were far more dependent on chemical engineering capabilities.

It must be stressed, however, that chemical engineering did have critical connections with science, although these connections were quite different from the usual portrayals. As chemical engineering matured, it created a framework that eventually made it possible to exploit scientific knowledge and scientific methodologies more effectively. But such possibilities were limited at the outset. It was only when the discipline of chemical engineering reached a reasonably mature stage that it became possible to exploit science in a more systematic way. The construction of that framework, or platform, was a necessary prior step. Engineering preceded science and laid the foundations that made the utilization of science possible.

Indeed, it is not stretching the GPT perspective excessively to say that the science of chemistry can be usefully thought of as an "applications field" of chemical engineering. We mean this in the specific sense that once the discipline of chemical engineering had been firmly established, the prospective economic payoff to chemical research of a more basic nature was substantially raised. Consider polymer chemistry, a field that was for many years dominated by research in private industry. It is doubtful that du Pont would have financed the fundamental researches of Carothers that led to the development of nylon were it not for the prior advances in chemical engineering that strengthened the firm's confidence that it could produce such new materials on a commercial scale.[3] The point is a more general one: Progress in the realm of engineering may provide powerful inducements to the performance of scientific research at a more fundamental level.

The design and construction of plants devoted to large-scale chemical processing activities involve an entirely different set of activities and capabilities than those that generated the new chemical entities. To begin with, such activities as mixing, heating, and contaminant control, which can be carried out with great precision in the lab, are immensely more difficult to handle in large-scale operations, especially if high degrees of precision are required. Furthermore economic considerations must obviously play a critical role in the design process. Cost considerations become decisive in an industrial context, and cost considerations are intimately connected to decisions concerning optimal scale of plant. An interesting feature of chemical engineering is that its practitioners have been much more deeply involved in dealing with cost considerations than other engineering professions.

Thus the discovery of a new chemical entity has commonly posed an entirely new question, one that is remote from the scientific context of the laboratory: How does one go about producing it? A chemical process plant is far from a scaled-up version of the original laboratory equipment. A simple, multiple enlargement of the dimensions of small-scale experimental equipment would be likely to yield disastrous results. Experimental equipment may have been made of glass or porcelain. A manufacturing plant will almost certainly have to be constructed of very different materials. Producing by the ton is very different from producing by the ounce. This indeed is what accounts for the unique importance of the pilot plant, which may be thought of as a device for translating the findings of laboratory research into a technically feasible and economically efficient large-scale production process. The translation, however, requires competences that are unlikely to exist at the experimental research level: a knowledge of mechanical engineering and physics and an understanding of the underlying economics of likely alternative engineering approaches.

Pilot plants, which in recent years have often been partially replaced by computer simulation exercises, have been indispensable in attempting to predict the performance of a full-scale production plant. Pilot plants are themselves a technology whose purpose is the reduction of uncertainties

---

3. For an authoritative treatment of the story of polymer research at Du Pont, see David A. Hounshell and John Kenly Smith, *Science and Corporate Strategy*, Cambridge University Press, 1988.

inherent in moving to large-scale production. Until a pilot plant is built, the precise characteristics of the output cannot be determined, and such essential activities as test marketing cannot proceed without the availability of reliable samples. Various other features of the production process cannot possibly be derived from any form of scientific knowledge alone. Consider the recycle problem. Very few chemical reactions are complete in the reaction stage. Therefore products of the reaction stage will not only include desired end products but also intermediates, unreacted feed, and trace impurities—some measurable and some unmeasurable. Impurities, in particular, are identified by the operation of the pilot plant in order to achieve a continuing steady-state condition.

It has been true of many of the most important new materials that have been introduced in the twentieth century that a gap of several, or even many years, has separated their discovery under laboratory conditions from the industrial capability to manufacture them on a commercial basis. This was true of the first polymers that W. H. Carothers produced with his glass equipment at the du Pont Laboratories. It was also true of polyethylene, one of the most widely used new materials of the twentieth century, and terephthalic acid, an essential material in the production of terylene, a major synthetic fiber. It was to manage such transitions from test tubes to manufacture, where output was to be measured in tons rather than ounces, than an entirely new methodology totally distinct from the science of chemistry began to be devised early in the century. The new methodology involved exploiting the central concept of "unit operations." This term, coined by Arthur D. Little at the Massachusetts Institute of Technology in 1915, provided the conceptual basis for what was to become a rigorous, quantitative approach to large-scale chemical manufacturing, and thus it may be taken to mark the emergence of chemical engineering as a unique discipline, in no way reducible to so-called applied chemistry.

Why is it incorrect to think of technologies as simply the legitimate offspring of scientific disciplines? Scientific progress over the centuries has proceeded precisely by abstracting from problems that are at the heart of the concerns of the engineering disciplines. Physics has, for example, like other scientific disciplines made great progress historically by confining itself to problems for which it can offer rigorous mathematical treatment and solutions. This has entailed abstracting from phenomena for which conceptual generality or rigorous solutions may be impossible. Thus, in formulating the classical laws of physics, Newton ignored the effects of friction in determining the acceleration of a falling object. Engineers, however, do not have the luxury of ignoring friction, viscosity, or turbulence when designing new machinery or equipment. They have had to develop designs that solve problems for which no scientific theory exists to provide guidance. Of necessity they have therefore taken to experimentation with small-scale working models, and at times they have had to confront awesomely complex problems in inferring the behavior of a projected full-scale prototype from the experimental performance of an observed, small-scale working model. How, in particular, does one adjust for scale effects? This is perhaps the single most persistent challenge in chemical process plant design

The main point is that the engineer's solution to the need for design data does not draw directly upon a scientific model, nor does the engineer's solution illuminate some more fundamental features of the phenomena under investigation. Nevertheless, the empirically derived coefficients are adequate for the design of industrial equipment that performs the required functions with a high degree of predictability and reliability. Importantly, engineering data developed by empirical experimentation can be used for generating design data for the prototype size by making use of dimensionless numbers. The best-known dimensionless number is the Reynolds number, with which it is possible to predict the transition from laminar to turbulent flow in an enclosed pipe or, later, for a moving object immersed in a fluid. Osborne Reynolds formulated this relationship in 1883 as a result of some ingenious experiments, and it has remained fundamental to the design of chemical process plants to this day.[4]

This statement, though made in the context of chemical engineering, would apply equally well in aeronautical engineering. In the hands of Ludwig Prandtl in Germany in the opening decade of the twentieth century, the Reynolds number provided the basis for the boundary layer concept that remains basic to the design of aircraft and thus made possible the calculation of drag on an airplane in flight. America's early successes, and eventual technological leadership in the aircraft industry in the first half of the century, owed a great deal to the aircraft design data generated at wind tunnels operated by the predecessor of NASA, the National Advisory Committee on Aeronautics, established in 1915. An additional contribution of empirical data, of great technological and commercial sig-

---

4. Reynolds "... showed that there is a critical velocity, depending upon the kinematic viscosity, the diameter of the pipe, and a physical constant (the Reynolds number) for the fluid at which a transition between the two types of flow will occur." *Dictionary of Scientific Biography*. vol. 11, p. 393. See also the paper by Edwin Layton Jr., The dimensional revolution: The new relations between theory and experiment in engineering in the age of Michelson, in Stanley Goldberg and Roger H. Stuewer, eds., *The Michelson Era in American Science. 1870–1930*, American Institute of Physics, 1988.

nificance, emanated from an exhaustive experimental study of the optimal shape of aircraft propellers. This study, carried out by William F. Durand and Everett P. Lesley, two Stanford professors of mechanical engineering, over the ten-year period 1916 to 1926, also relied heavily on empirical data drawn from experiments with wind tunnels and associated equipment. In assessing the real significance of these data, it is important to observe that the findings did not actually lead to drastic changes in propeller design. Rather, they accomplished something of much more general significance, or perhaps one should say of much more "general purpose." "The true significance of the Durand-Lesley model data was that they greatly improved the airplane designer's ability to match the propeller to the engine and airframe. This improvement was reinforced by the guidance available from the comparison between model and full-scale results and by the ways of thinking that had been developed. The designer now had at hand much needed tools for the improvement of airplane performance."[5]

7.3   Emergence of Chemical Engineering

Where did chemical engineering first emerge? As an academic discipline the answer is clearly at the Massachusetts Institute of Technology in the second decade of the twentieth century. In fact MIT's early intellectual leadership here was so dominant that the following discussion will focus exclusively on that institution. Of course there were antecedents to which chemical engineering can be traced. In England, George Davis had given a series of lectures in Manchester in 1887 that were later developed into a book published in 1901, *Handbook of Chemical Engineering*, but Davis made very little impact in England. As early as 1888 Lewis Mills Norton had taught a course in the Department of Chemistry at MIT that was labeled "chemical engineering." But the course mainly consisted of lectures that described the commercial manufacture of chemicals used in industry and were based largely on German practice.[6] Norton's lectures offered little treatment of the mechanical engineering aspects of the design of large-scale chemical process plants, and so they were more deserving of the term "industrial chemistry." Industrial chemists at the time focused on the production of a very large number of chemical products. They were concerned with sequences of steps, from beginning to end, in the production of individual products and not with the unifying principles

---

5. Walter Vincenti, *What Engineers Know and How They Know It*. Johns Hopkins University Press, 1990, p. 158.
6. *The Improbable Achievement: Chemical Engineering at MIT* by Harold Weber, MIT Press, 1979, p. 7.

between the manufacture of different products, so few such principles were identified.

It was not until 1915 that an original unifying concept of this discipline was forcefully articulated. This was the "unit operations" concept of Arthur D. Little. Little first presented this concept in a report to MIT's Corporation in December 1915 as support for the establishment of a School of Chemical Engineering Practice at the Institute. In Little's words: "Any chemical process, on whatever scale conducted, may be resolved into a coordinated series of what may be termed 'unit actions,' as pulverizing, mixing, heating, roasting, absorbing, condensing, lixiviating, precipitating, crystallizing, filtering, dissolving, electrolyzing and so on. The number of these basic unit operations is not very large and relatively few of them are involved in any particular process.... Chemical engineering research ... is directed toward the improvement, control and better coordination of these unit operations and the selection or development of the equipment in which they are carried out. It is obviously concerned with the testing and the provision of materials of construction which shall function safely, resist corrosion, and withstand the indicated condition of temperature and pressure...."[7]

A critical feature of this concept of unit operations is that it went far beyond the mere descriptive approach of industrial chemistry by calling attention to a critical few distinctive processes that were common to a number of industries. This act of intellectual abstraction, which Little initiated, laid the foundations for a more rigorous and, eventually, more quantitative discipline. Little looked at the large number of unique vertical sequences which had until then described the manufacturing steps of individual chemical products and then looked across these sequences, horizontally, to draw together the small number of common elements in each of them. It seems reasonable to call Little's statement an attempt to provide a general purpose technology to industries that made use of chemical processing.

Furthermore such chemical processing plants existed in many industries besides the chemical industry proper. Some of the larger ones included petroleum refining, rubber, leather, coal (by-product distillation plants), food-processing, sugar refining, explosives, ceramics and glass, paper and pulp, cement, and metallurgical industries (e.g., aluminum, iron, and steel).[8] Clearly, a critical part of the case for regarding chemical engineering as a

---

7. Arthur D. Little, *Twenty-five Years of Chemical Engineering Progress*, 1933, American Institute of Chemical Engineers, pp. 7–8.
8. See the table on the distribution of chemical engineering graduates, 1927–29, in *Chemical and Metallurgical Engineering*, July 1929, p. 415.

general purpose technology is that its impact has been felt far beyond the limits of the traditional chemical industry, even before the postwar rise of the petrochemical industries. In fact Arthur D. Little had run a consulting firm for many years and was therefore well-positioned to make this comparative judgment on manufacturing technologies. His main professional commitment was to the paper and pulp industry in which he had consulted widely, and as a young man he had worked for the first mill in the United States that employed the new sulfite process for converting wood pulp into paper.[9] That experience was to play a significant role in Little's connections with MIT, where he lectured on papermaking from 1893 to 1916. It seems reasonable to conclude that Little's consulting experience, concentrated as it was in industries where chemical transformations were central to the manufacturing process, sharpened his awareness of the extent to which entirely different product lines were making use of similar operations.

The concept of unit operations put chemical engineering in a position to accumulate a set of methodological tools that provided the basis for a wide range of activities connected with the design of chemical process plants. The awareness that there were a limited number of similar operations common to many industries served to identify research priorities and therefore to point to a disciplinary research agenda. This was not possible so long as technologists remained lost in the particularities of individual products. For example, a widely used British reference book for technical workers in the midnineteenth century began its article on "Distillation" as follows: "Distillation means, in the commercial language of this country, the manufacture of intoxicating spirits; under which are comprehended the four processes, of mashing the vegetable materials, cooling the worts, exciting the vinous fermentation, and separating by a peculiar vessel called a still, the alcohol combined with more or less water."[10] Tunnel vision with respect to an operation as common as distillation was obviously incompatible with the notion of a chemical engineer who dealt with a limited number of operations common to a wide range of industries. It is no doubt true that some people in Britain knew of distillation only in the context of breweries, but it is surprising that this included the author of an article in a reference book for technical workers.

Moreover the concept of unit operations had great pedagogical value; it provided the basis for a curriculum that could be taught. All this did not happen immediately, however. The establishment of a curriculum of mar-

---

9. See E. J. Kahn Jr., *The Problem Solvers*, Little, Brown, 1986, p. 25.
10. Andrew Ure, *Dictionary of the Arts. Manufactures and Mining* 1853, as quoted by W. K. Lewis, Evolution of the unit operations, *Chemical Engineering Progress*, symposium series, 1955, p. 2.

ketable skills had to await more extensive interaction between universities and private industry. At MIT, in 1920, just a few years after Little's formulation, chemical engineering achieved the status of a separate department under the chairmanship of W. K. Lewis. For the next few decades the teaching of chemical engineering was organized around the concept of unit operations, though the concept underwent substantial alteration in its intellectual content almost from the beginning.

An engineer trained in unit operations could mix and match these operations and be more flexible and resourceful in his approach to problem solving. Most important, he was equipped to take the methods from one area of industry and transfer them to other areas that were unrelated in terms of final products.[11] This capability was especially valuable in the innovation process—particularly as new materials and new intermediate products emerged. Thus research that improved the efficiency of any one process was now likely to be more quickly employed in a large number of places. Putting the point somewhat differently, the identification of a small number of unit operations common to a large number of industries meant that it was now possible to identify specific research topics where new findings could be confidently expected to experience widespread utilization.

The way in which chemical engineering was institutionalized in the United States meant that the university was the locus of research. This ensured that the focus of intellectual progress would be on general results rather than merely the ad hoc solutions of separate industrial problems. The university maintained strong links with private industry especially for funding, and this further ensured that university research would remain focused on issues of direct relevance to industry needs. In fact one could speculate that the overwhelming U.S. leadership in chemical engineering has owed much to the close connections between university and industry. A much more prominent role of the government as a funder of research, as was the case in Europe, would have led to a very different outcome.[12]

---

11. See Peter H. Spitz, *Petrochemicals*, Wiley, 1988, pp. 58–59.

12. One eminent chemical engineer offered the following retrospective observations: "Especially in earlier times, research ideas usually were found in the uncertainties that arose in the course of attempts to design equipment, to scale up reactors, to develop economical and efficient methods of manufacture, or to evaluate proposed processes. Even in university laboratories, research subjects were chosen with the probable use of the results in chemical manufacture as the main object. Because of the educational focus on unit operations rather than on particular processes, however, research results were sought that would have general significance. Correlations of experimental data in their most widely applicable form were common and much admired. Some results of this kind, obtained during the period from 1930 to 1960, when unit operations work was most popular, are still in use today with little, if any, modification." Robert Pigford, Chemical technology: The past 100 years, *Chemical and Engineering News*, April 6, 1976, p. 197.

7.4 Maturing and Expanding Application of Chemical Engineering

In the interwar years and post–World War II years, the discipline of chemical engineering underwent a huge expansion in its industrial coverage and impact. This expansion was made possible by the fruitfulness of the unit operations approach—or more precisely, by the transformation of a rather loosely defined concept into one of steadily increasing quantitative rigor.

In the years right after Little's formulation of the concept of unit operations, research in chemical engineering sought to acquire a deeper understanding of each operation in order to establish mathematical regularities that could reduce the cost of the designing process as well as improve the efficiency of the equipment being designed. Eventually these separate operations were reduced to more inclusive concepts such as fluid mechanics and heat transfer. These concepts became more inclusive and incorporated processes based on the principles underlying the operation of gas, liquid, or solid components in different concentrations.[13]

In time chemical engineers attempted to codify the basic physical underlying momentum (viscous flow), energy transport (heat conduction, convection, and radiation), and mass transport (diffusion).[14] It was continual advancements like these that have led recently to the claim that chemical engineering has achieved the status of an "engineering science." As a result much of the designing activity of chemical engineers is now understood at a more fundamental level, while the design of chemical processing equipment still draws upon empirical regularities that have stood the test of time.

But the success, and indeed the transformation of the discipline, owed an enormous debt to two exogenous events. The first was World War I which brought in its wake a great expansion in the demand for chemical engineers to supply the munitions, nitrates, gasoline, and other requirements of a wartime economy (including chemical warfare).[15] In the United States the number of students attending chemical engineering courses, which had remained far behind those in electrical engineering and mechanical engineering before the war, expanded rapidly in the second decade of the century, growing by more than 60 percent between 1916 and 1918 alone, and rising to almost 6,000 in 1920 to 1921.[16]

---

13. F. J. van Antwerpen, in John McKetta, ed., *Encyclopedia of Chemical Processing and Design*, vol. 6, New York, 1978, p. 335.
14. van Antwerpen, ibid., p. 356.
15. See W. Haynes, *American Chemical Industry: A History*, Van Nostrand, 1945, vol. 2, for details.
16. See L. F. Haber, *The Chemical Industry 1900–1930*, Oxford University Press, 1971, p. 63; Terry Reynolds, *75 Years of Progress: A History of the American Institute of Chemical Engineers*, American Institute of Chemical Engineers, New York, 1983, p. 12.

The second, and more enduring, exogenous event was the spectacular growth in the automobile industry, with its voracious appetite for liquid fuel. There were fewer than 500,000 registered cars in the United States in 1910, but over 8,000,000 in 1920, and over 23,000,000 in 1930, by which time the automobile industry was the largest manufacturing industry, in terms of value added, in the United States[17] After 1920 the history of chemical engineering simply became inseparable from the history of petroleum refining. The new technologies and new methodological skills developed by the chemical engineering profession in satisfying the demand for gasoline eventually had enormous unanticipated consequences. The usefulness of these capabilities multiplied when, after World War II, first the U.S. chemical industry and then the world chemical industry shifted their resource base to petroleum feedstocks, creating the present-day petrochemical industry. The United States had substantial commercial advantages in the petrochemical industry. Not only were there large domestic deposits of petroleum, but this country had earlier traversed petroleum refining and built up chemical engineering capabilities in that activity that later flowed from petroleum refining to petrochemicals. Thus resource endowment mattered, but it mattered in large part because it served as a powerful stimulus to acquire technological capabilities that simply did not exist in 1920. This early experience in petroleum refining served to provide powerful first-mover learning advantages for the United States when the world later turned to petrochemicals.

With that transition, the technologies that had been specifically developed for the refining of petroleum eventually came to provide the technological basis for a much larger share of the world's industrial output. Skills that were initially acquired in petroleum refining were later transferred to the much larger canvas of the emerging petrochemical industry, including major new product categories such as plastics, synthetic fibers, and synthetic rubber. The "crash" program to develop synthetic rubber during World War II was almost entirely the achievement of chemical engineers. But many of the new capabilities spilled over into inorganic chemical products as well. The new technology and its broader design, along with new problem-solving skills, had an impact well beyond the chemical industry. In addition to the beneficiaries mentioned earlier, chemical engineers eventually became responsible for the processing of uranium for nuclear power plants. Their development of the manufacturing technology of submerged fermentation also made the manufacture of penicillin possible during the war (see below).

---

17. *Historical Statistics of the US*, U.S.G.P.D., 1975.

Thus, in attacking the problems of large-scale refining of petroleum, chemical engineers created a vastly expanded pool of design and problem-solving capabilities that were critical to the creation of some entirely new industries. This further made possible a shift from batch production to large-volume production methods and introduced automatic controls to continuous processing in a number of industries. At the same time, it also made a substantial contribution to the whole realm of automatic control. Although continuous processing plants already existed in certain places, automatic controls raised considerably the productivity of continuous processing plants. For example, continuous cracking technology, which was introduced in the 1920s, could achieve far higher levels of throughputs when the continuous cracking process plants were operated with automatic controls.[18]

Seen from the perspective of the 1990s, the close connection between the petroleum and chemical industries seems natural and inevitable. But this was not the vantage point of the 1920s. The two industries were in reality brought together by a complex human creation involving the mobilization of vast amounts of resources and the sustained exercise of human intelligence in developing inventions, novel designs, and wide-ranging new problem-solving methodologies. Before the invention of the internal combustion engine, petroleum was valued primarily as an illuminant and a lubricant, and the more volatile fractions were commonly disposed of as waste. It became a major source of fuel for transportation purposes only as a result of the invention of the automobile. The technologies that chemical engineering developed, in order to extract fuel from petroleum, provided much of the technological basis for a far larger group of industries—petrochemical industries—in the postwar years. Indeed, to emphasize once more, the new technological capabilities spilled over into many sectors outside of both petrochemicals and the older chemical industries.

In 1920 the state of technological knowledge was such that petroleum was not regarded as a significant input into the chemical industry. The chemical industry thought of its inputs in terms of chemicals in a less processed state, on the one hand, and feedstocks drawn from coke-oven by-products, on the other. It was only as the result of extensive research and the slow accumulation of technical knowledge that the oil companies

---

18. See, for example, John H. Lorant, *The Role of Capital-Improving Innovations in American Manufacturing during the 1920s*, Ph.D. dissertation, Columbia University, 1966. p. 103; George Perazich et al., *Industrial Instruments and Changing Technology*, WPA, Philadelphia, 1938, pp. 66–67.

came to realize that their refining operations could produce not just fuel and lubricants but organic chemical intermediates as well.[19] In the 1920s the offgases of oil refineries, if they were not simply flared, were likely to be employed only as fuel, a low-value use, at the refineries themselves. The transformation of the chemical industry into the petrochemical industry that matured in the post–World War II years was in large measure an achievement in which by-products that were formerly treated as waste materials were converted into sources of great commercial value.[20] In a sense it was a replay of an earlier development in nineteenth-century Germany when coal tar, a by-product derived from coke ovens, also provided the raw material basis for a burgeoning synthetic dye industry. In fact it would be no exaggeration to say that Standard Oil (New Jersey) and Shell were induced to undertake the research that brought them from petroleum refining into the chemical industry by their growing awareness of the commercial opportunities that might flow from the eventual utilization of the waste products of their refinery operations.[21]

7.5 Role of MIT

The achievement of chemical engineering as a discipline involved a uniquely intimate set of interactions between the petroleum refining industry and the newly formed chemical engineering department at MIT. During the 1920s that department built upon the conceptual platform of unit operations that had been introduced by Arthur D. Little. But that concept needs to be understood as the inevitable crude starting point of the discipline of chemical engineering and not its terminus. In fact the concept was enriched and deepened very quickly as MIT's chemical engineers confronted the difficulties of satisfying the demand for gasoline. The technique of thermal cracking was first introduced by Standard Oil in 1913 (the Burton process). Thermal cracking was far cheaper than the "straight-run" distillation method that it replaced ("straight-run" refers to

---

19. See Spitz, op. cit. ch. 2.
20. For example, the origin of the central research laboratory for the Royal Dutch Shell Oil Company's American subsidiary has been attributed to the growing awareness that there were huge amounts of oil field and refinery gases that had been previously flared or merely burned for boiler fuel. The importance of developing a research program to utilize these by-products was closely linked, in turn, to the growth in scale of plant: "The quantities of these gases were at first insignificant, but with rapid expansion of cracking facilities in the second half of the 'Twenties, the volume of cracking gases became enormous." K. Beaton, *Enterprise in Oil: A History of Shell in the United States*, Appleton-Century-Crofts, 1957, pp. 502–503, as quoted in David Mowery, *The Emergence and Growth of Industrial Research in American Manufacturing, 1899–1945*, Ph.D. dissertation, Stanford University, 1981, p. 123.
21. Spitz, op. cit., pp. 68, 116, and 514–15.

the refinery streams obtained from the fractional distillation process), and it was to remain dominant in petroleum refining until just before the Second World War. Thermal cracking was analyzed with increasing sophistication, in terms of its specific unit operations, including heat transfer, fluid flow, and distillation.[22] The design process was approached in increasingly quantitative rather than crudely empirical terms. In the early 1930s, for example, pressure-volume-temperature relationships for gas mixtures were established that were sufficiently accurate for the design of refining equipment, and this knowledge proved to be useful elsewhere.[23]

Undoubtedly the leading academic contributor was W. K. Lewis, the chairman of MIT's chemical engineering department. Lewis served as a consultant to Standard Oil of New Jersey (later Exxon) for many years (Exxon Research and Engineering had been founded in 1919). His first efforts at Standard Oil were to provide precision distillation equipment and to convert batch-processing methods to methods that were both continuous and automatic, as in thermal cracking (the tube-and-tank process) and continuous vacuum distillation. In doing these things, he was also, unknowingly of course, inventing technologies upon which the petrochemical industries of the future were to be based.

Distillation is the most important single activity in petroleum refining. Lewis and his MIT colleagues during the 1930s "... advanced the theory of fractional distillation to the point at which equipment could be designed that would split multicomponent hydrocarbon streams predictably and consistently into the desired fractions."[24] By 1924 Lewis had helped to achieve a significant increase in oil recovery by the use of vacuum stills. (Between 1914 and 1927 the average yield of gasoline rose from 18 to 36 percent of crude throughput.) This work and his earlier bubble tower designs became refinery standards.

At the same time course work at MIT was quickly expanded to embody these new concepts and their underlying design principles. Lewis, along with two colleagues, William Walker and William McAdams, published the first edition of what was to be an immensely influential textbook, *Principles of Chemical Engineering*, in 1923. It was a text that took a long step forward from A. D. Little's initial formulation of unit operations to a methodology of far greater quantitative and mathematical rigor. In this respect it provided a much-improved platform for the introduction into chemical engineering research of scientific concepts such

---

22. Continuous cracking processes accounted for about 90 percent of cracked gasoline capacity in 1929. See Lorant, op. cit., table B-10.
23. *Improbable Achievement*, op. cit., p. 28.
24. ACS, *Chemistry in the Economy*, p. 282.

as thermodynamics. Indeed thermodynamics and transport phenomena are now engineering concepts at the heart of the discipline of chemical engineering.

In the late 1920s Exxon negotiated a series of agreements with the German chemical giant, IG Farben, that would provide access to their extensive research on hydrogenation and synthetic substitutes for oil and rubber from coal. It was also anticipated that the German findings might be used to increase gasoline yields and promote Exxon's entry into chemicals. In putting together a new research group for this purpose, Lewis was again consulted. He recommended Robert Haslam, head of the MIT Chemical Engineering Practice School. Haslam, on leave from MIT, formed a team of fifteen MIT staff members and graduates who set up a research organization in Baton Rouge, Louisiana. Many of the members of this group later rose to positions of eminence in petroleum and chemicals.[25]

Much of what subsequently took place in modern petroleum processing until the Second World War originated in Baton Rouge, and the primary responsibility for these achievements was the solid phalanx of MIT chemical engineers who worked there. With the continuing advice of Lewis, and later (1935) of MIT Professor Edwin R. Gilliland, Baton Rouge produced such outstanding process developments as hydroforming; fluid flex coking and fluidized bed catalytic cracking were also introduced. This last innovation was a contribution to petroleum refining that was to play a big role in World War II, and in the postwar world it dominated the refining of petroleum and ultimately became the most important processing technology for propylene and butane feedstocks in the chemical industry. Catalytic cracking plants provided the essential raw materials for the "crash" program for the production of synthetic rubber in World War 11. They also provided the high-octane gasoline that significantly improved allied fighter aircraft performance. The key patent was originally applied for by Professors W. K. Lewis and E. R. Gilliland in January 1940 and assigned to Standard Oil Development Company.[26]

Lewis's strategy, as a chemical engineer, was that of an overall systems approach to the design of continuous automated processing plants for the refining of petroleum. This approach, and many of the specific design

---

25. Ralph Landau and Nathan Rosenberg, Successful commercialization in the chemical process industries, in Nathan Rosenberg, Ralph Landau, and David Mowery, eds., *Technology and the Wealth of Nations*, Stanford University Press, 1992, ch. 4.

26. See John Enos, *Petroleum: Progress and Profits*, MIT Press, 1962, pp. 196–201. Hydroforming is "a process for improving the octane number of low grade virgin gasoline." Fluid coking is "a method for converting heavy residual crude fractions into higher value, lighter boiling fractions." E. J. Gohr, Background, history and future of fluidization, in Donald Othmer, ed., *Fluidization*, Reinhold Publishing, 1956, p. 115.

ingredients, shaped much of the later technology of the world petrochemical industry. Petroleum refining thus provided vitally important learning experiences in the design of the continuous flow, automated processing technology that was later transferred to the much more diversified canvas of petrochemicals in the postwar years—plastics, synthetic fibers, synthetic rubber.[27]

It is important to stress the close relationship between the consulting activities of the chemical engineering faculty and the training of future chemical engineers at MIT. Faculty encountered research problems in their consulting activities and brought these problems back to the Institute where students might pursue these research problems under faculty supervision. Problems encountered in the course of faculty consulting were likely to appear on their lists of suggested thesis topics.[28] At least on some occasions the results were momentous for industry.

Consider the development of fluidized bed catalytic cracking. Fluidized bed technology was just one of the ways in which chemical engineers rendered chemical processing continuous. But it was probably the most important, single continuous process innovation. Petroleum "cracking" techniques make it possible to break large, heavy hydrocarbon molecules into smaller and lighter ones. Catalytic cracking technology made possible a much higher degree of control over the output that could be extracted from a given quantity of petroleum. In practice, this meant raising the yield of higher-priced gasoline that could be derived from the heavier crude oil fractions, an achievement of enormous economic significance.

Although Houdry had introduced the use of a catalyst in petroleum refining in the late 1930s, his fixed bed catalytic process, while a major breakthrough, suffered from a serious limitation. It deposited carbon on the catalyst, leading to rapid degeneration, and required regeneration. (The depositing of carbon led to severe deactivation in ten or twenty minutes.) In short, the Houdry process was not continuous. There was also a very different, but far from trivial problem: Houdry was asking for a very high licensing fee.

Although the Houdry process was already being commercialized in the late 1930s, Exxon (then Standard Oil company of New Jersey) decided to

---

27. On the relevance of this learning to sectors outside petrochemicals, such as paper and pulp, uranium processing, foodstuffs, and others, see Spitz, op. cit., pp. 135–38 and *The Encyclopedia Britannica*. pp. 124–25. See also Edward Gornowski, The history of chemical engineering at Exxon, in Furter, op. cit.
28. Spitz, op. cit., p. 133.

search for the development of a more efficient process of their own. After numerous engineering studies, it was concluded that there were inherent and insuperable limitations in a cyclic, fixed-bed design. Consequently the focus of research was shifted to circulating catalytic systems that would permit continuous operation.[29]

In 1938 W. K. Lewis and E. R. Gilliland in their roles as consultants to Standard Oil suggested that the tube in which the reaction took place should be vertical rather than horizontal. To investigate the behavior of finely divided particles in vertical tubes, they initiated a research program at MIT. Experiments carried out by two graduate students, John Chambers and Scott Walker, succeeded in deriving some of the basic engineering relationships underlying the fluid technique. They worked with air and catalyst mixtures alone. But, as Standard Oil immediately realized, the fluid technique offered an excellent mechanism for manipulating the catalyst and oil streams, and they quickly took over the research. Within six months their laboratories were cracking oil in the presence of a fluid catalyst.[30]

Thus fluidized bed catalytic cracking was largely the joint achievement of MIT and the Esso laboratories. The purely theoretical work and laboratory-scale experiments were performed primarily at MIT, while the initial problem formulation and the later scaled-up pilot plant experiments were carried out by private industry. (Other contributors to fluid catalytic cracking included Universal Oil Products Company, Kellogg Company, Texas Development Corporation, Gulf Research Development Company, and Shell Oil Company.)

The principles of fluidization, developed in the specific context of petroleum refining, were subsequently to experience extensive applications in other types of chemical processing, not only in the newly emerging petrochemical industries but also over a much broader spectrum of chemical processing activities. The research carried out jointly between MIT and Exxon, on the flow properties of powdered solids suspended in gases, turned out to have a much larger potential relevance. "Potentially, the fluid technique has application to any process in which (1) large quantities of heat are transferred; (2) large quantities of solids must be circulated; or (3) very intimate contact between gases and solids is desired."[31] In fact the potential uses are so great as to constitute the introduction of an entirely new unit operation.

---

29. Jahnig et al., p. 276.
30. Enos, op. cit., pp. 200–201; see also p. 281.
31. Gohr.

By 1962 it was estimated that there were 350 fluidized-bed processing units throughout the world outside the petroleum industry.[32] At the same time, within the petroleum industry, a technique that had been developed for the specific purpose of catalytic cracking was also being used in fluid-bed coking, catalyst regeneration, platforming, and ethylene manufacture.[33]

A basic advantage of the fluidized-solids technique was the uniformity that it achieved with respect to particle size distribution, temperature, and heat transfer characteristics throughout the entire fluidized bed. This made it possible to established precise conditions of control comparable to what could presumbly be achieved before only in the laboratory. More important, it made possible continuous process activities for a wide variety of end uses in other industries. The technique also represented a vast improvement with respect to ease of handling.[34]

7.6 Rise of the Petrochemical Industry

The rise of the petrochemical industry began with the entry of a few companies, particularly Union Carbide, Shell, Dow, and Exxon, not long before the Second World War. These firms soon encountered the need for chemical engineering skills as the increasing scale of operations forced resort to continuous processing, just as it had previously in the petroleum refining industry. The techniques and skills formerly developed for the refining industry could now be applied to the demands of the petrochemical industry, as chemical raw materials began the epochal shift from coal-based to petroleum-based feedstocks. To be sure, it was not a simple shift. The problems in manufacturing chemicals were different, and in many respects even more challenging. They concerned corrosion, complex product separations and purifications, toxic wastes and hazards, and the like. But the prior experience with petroleum refining provided a vast storehouse of concepts, methodologies and, not least, experience, upon which the chemical engineering profession could draw. In several cases the early entrants were the oil companies going downstream to capture the newly available rents.

Although the episode is not well-known, the skills of the chemical engineer in the use of pilot plants and experimental design played a crucial role in the wartime introduction of penicillin. Even though Alexander

---

32. *Chemical Engineering*, July 9, 1962, p. 125.
33. Loc. cit.
34. *Chemical Engineering*, May 1953, p. 220, and July 9, 1962, p. 126.

Fleming's brilliant insight, that a common bread mold was responsible for the bactericidal effect that he observed in his Petri dish, was made in 1928, penicillin remained unavailable at the outbreak of the Second World War. Producing penicillin on a very large commercial scale during that war required a "crash program" in which the production problems were solved not, as might be expected, by the pharmaceutical chemist but by chemical engineers designing and operating a pilot plant. The chemical engineers demonstrated how the technique of aerobic submerged fermentation, which became the dominant production technology, could be made to work by solving the complex problems of heat and mass transfer.[35]

There is a revealing aspect to the wartime development of penicillin that is also worth observing. Although the British clearly pioneered in the scientific research leading to the development of penicillin, their main subsequent research interest was in finding new uses for penicillin and in improving its effectiveness in the clinical treatment of infection. In Germany, on the other hand, where there was a very strong tradition and accumulation of skills in chemical synthesis, the synthesis route was the preferred approach. As it was later established, this route was much more difficult and costly than obtaining the penicillin directly from the mold.[36] In America, by contrast, as soon as the significance of penicillin was fully appreciated, the skills of the chemical engineer were enlisted to identify efficient ways of achieving large-scale production methods and increased yields. The joint achievement of the chemical engineer and the microbiologist should perhaps best be described as the first great achievement of biochemical engineering.

The maturing of the discipline of chemical engineering also gave rise to an important organizational innovation that was to accelerate the diffusion of new chemical processing technologies: specialized engineering firms (SEFs). The skills in designing continuous-flow, automated processes, first acquired in petroleum refining, could be, and were, exploited worldwide in the postwar years. Since the development of new chemical products was largely in the hands of big firms which performed the necessary R&D, there existed an important niche that could be filled by small firms, consisting of well-trained chemical engineers, that could concentrate their efforts exclusively on the design of chemical process plants. A large petrochemical firm that designed its own equipment was severely limited

---

35. See AICHE, *The History of Penicillin Production*, American Institute of Chemical Engineers, 1970.
36. John C. Sheehan, *The Enchanted Ring: The Untold Story of Penicillin*. The Germans were also influenced by their expectation that sulfa would eventually prove to be the drug of choice.

in its ability to benefit from accumulating experience, since the need to do such designing for its own internal purposes was, at best, only intermittent. On the other hand, small specialized chemical engineering firms working for potentially large, indeed world, markets had numerous opportunities for improving both their designing and innovating capabilities by designing particular plants many times for a succession of different clients. Indeed the feedback of know-how from a succession of clients was a major source of technological learning and competitive advantage to the most successful specialized engineering firms, such as the Scientific Design Company.

Perhaps even more important was the rapid diffusion of recently acquired technological knowledge that resulted from the activities of the SEFs. These firms played a significant role in developing and licensing processing technologies, thus enabling newcomers such as Conoco, Arco, and Amoco to enter the petrochemical industry in the 1960s and 1970s, and to do so with, for example, ethylene plants that were at least as efficient as those already under operation by the traditional large producers.[37] In effect SEFs represented a new pattern of specialization in which certain firms thrived by becoming the vehicles for rapidly diffusing or "externalizing" the new technologies developed in petroleum refining.

7.7 Some Concluding Observations

In the 1920s a newly established chemical engineering discipline set about the business of designing more efficient processing technologies in order to satisfy the rapidly expanding demand for gasoline. In achieving this end, it appears to have also gone a long way toward developing the processing technologies that were to be essential to the vast expansion of the petrochemical industries, beginning around the time of the Second World War. In doing so, the discipline of chemical engineering became, in effect, a general purpose technology.

Of course chemical engineering did not become a general purpose technology entirely through its own actions. Although this chapter has emphasized the growing needs of the automobile in stimulating the growth of chemical engineering, its expansion required the later growth of the downstream industries that would expand the range of productive operations to which the technologies of the chemical engineer might be applied. In large measure chemical engineering became a GPT when research in polymer chemistry (Staudinger, Meyer, Mark, Carothers) laid

---

37. Spitz, op. cit., pp. 320, 424, and 456–58.

the scientific basis for a whole range of new products that could best be produced from petroleum feedstocks. The scientific basis for these new products that were to dominate the petrochemical industry—plastics, synthetic fiber, synthetic rubber—was also, like chemical engineering, a product of the interwar years. But these petrochemical industries only acquired major economic significance in the years after World War II.

This interesting, serendipitous feature of chemical engineering as a general purpose technology did not emerge in response to the prior existence of a large, heterogeneous body of downstream users of the technology, whose existence generated strong economic incentives to upstream suppliers of complementary inputs. Complementarities were, indeed, important, but the story of chemical engineering's general purpose capability was developed primarily in response to the needs of one sector: petroleum. Only later did petroleum come to constitute the main feedstock for an expanding multiplicity of industrial users. Thus developments upstream in the interwar years clearly preceded the extensive developments downstream that eventually transformed chemical engineering into a GPT. The story is one of large unanticipated benefits in which a capability originally developed for one limited set of needs turned out to satisfy a much larger set of needs.

A critical achievement of the initial concept of unit operations was that it clarified the objectives of research. In this sense, it served as a focusing device. By providing an intellectual platform for science, it eventually altered the nature of the platform itself. Thus chemical engineering did not emerge out of prior science but rather, as it matured, strengthened the opportunities for focusing scientific concepts and methodologies upon the problems with which it dealt. As a result chemical engineering eventually *became* more scientific.

This chapter has emphasized the crucial role that one institution of higher education, MIT, played in the rise of chemical engineering. Some further aspects of the impact of that role ought to be made more explicit. As we have seen, the interface between MIT and private industry was a very intimate one, with professors serving as regular consultants over extended periods of time in dealing with the problems of petroleum refining. So long as professors maintained an active role in a teaching capacity, they were under a natural pressure to place the knowledge, acquired from their specific problem-solving activities as consultants, in a larger and more general context. This meant fitting that knowledge together in an internally consistent way with other knowledge in their discipline. In brief, when reverting to their teaching roles, they needed to systematize their knowledge, a natural and essential precondition for the writing of textbooks as well as other forms of publication. This had profound implications for the diffusion of new technological knowledge, not just because open universities "naturally" diffuse their knowledge but because the need to systematize knowledge for teaching purposes meant that they had to spend time and sustained effort in further activities that inevitably facilitated the spread of useful knowledge.

The story recounted here is not reducible to a simple model of human agents driven by maximizing behavior. Interestingly it involves the ways in which an institutional location of one group of actors—university professors—led to a more complex pattern of behavior. It was the setting, and not just maximizing behavior, that brought about the emergence of a new academic discipline. The growing demand for refined petroleum products in turn led to the training of an enlarged cohort of chemical engineers who shaped the creation of a huge petrochemical industry. An academic discipline therefore generated not only the expected "vertical" externalities of lower downstream invention costs; it also generated a "horizontal" externality.

To give this argument much larger scope, it may be said that the petrochemical industry was in fact more the achievement of Du Pont than of Exxon. Exxon "merely" increased the supply and reduced the price of petroleum, whereas du Pont employed the discipline of chemical engineering to introduce and to produce a vastly expanded array of new products that utilized the cheaper petroleum inputs.[38] Nevertheless, a prominent role was played by an academic institution devoted to teaching as well as research. All students, whatever their eventual professional affiliation, were taught the common language of chemical engineering. Even though they went to work in different firms or organizations, they had common concepts, theories, and methods on which to build their work. This vastly reduced the barriers to the diffusion of technical knowledge across organizational boundary lines. It therefore facilitated the development of a professional community of people who could communicate easily with one another. Indeed, this would appear to be a peculiar feature of a *discipline* as a GPT, as opposed to a piece of *hardware* as a GPT. It was what made chemical engineers, after the introduction of the concept of unit operations, so different from the earlier industrial chemists, who tended to speak in very idiosyncratic, industry-specific, or even

---

38. Du Pont was also responsible for considerable contributions to the expanding knowledge base that eventually made chemical engineering a more sophisticated discipline. See Hounshell and Smith, op. cit., ch. 14 (see also note 3 above).

firm-specific, languages. To some extent, this would appear to be what happened in Germany, where chemical engineering did not emerge as a distinct academic subject area until after the Second World War, when it was essentially borrowed from America. In Germany, universities played an important role in training students in chemistry, but these chemists had no engineering skills and therefore required extensive on-the-job training in the different chemical divisions of the firms in which they were eventually employed. It seems likely that this led to more secrecy and less extensive interfirm communication in Germany.[39]

The notion of a general purpose technology has, up until now, been confined to machines or devices that are utilized over some wide range of productive activities. We suggest that a broadening of the concept to include intellectual methodologies, such as in chemical engineering, may bring with it valuable new insights into the underlying determinants of technological change and the diffusion of new technologies. As was suggested at the beginning of this chapter, a distinctive feature of the second industrial revolution was the emergence of new engineering disciplines that have become the centers of technological learning and the carriers of technological change to their respective industries. In this sense, studying the development and functioning of these disciplines offers the enticing prospect of penetrating to a deeper level of understanding of the technological dynamics of industrial economies. It may even vastly enlighten us on what is the new growth theory's most critical and perhaps weakest link: accounting for the growth of useful knowledge.

---

39. See, for example, J. C. Guedon, Conceptual and institutional obstacles to the emergence of unit operations in Europe, in William Furter, ed., *History of Chemical Engineering*, American Chemical Society, New York, 1980, pp. 67–68; Ralph Landau, Chemical engineering in West Germany, *Chemical Engineering Progress*, July 1958.

8   The Consequences of Changes in GPTs

Richard G. Lipsey, Cliff Bekar, and Kenneth Carlaw

In this chapter we focus on the consequences of the evolution of general purpose technologies, specifically the replacement of established technologies by new GPTs.[1] As in chapter 2, we present a set of facts that we hope will help constrain theorizing in productive ways. In section 8.1 we lay out a model that distinguishes between technology and the economy's structure. Although this model is not necessary for defining and identifying GPTs, it helps to organize our knowledge and to theorise about their effects. In section 8.2 we study the effects that new GPTs have on other technologies, on what we call the facilitating structure, and public policy. In section 8.3 we look at how these effects influence economic performance. We conclude in section 8.4 with some challenges for future theorising that are suggested by our analysis.

8.1 A Structuralist Model

In this section we outline an appreciative theory designed to organize existing empirical knowledge about changes in technologies. To isolate the various effects, the theory separates technology from its embodiment. We summarize its elements and contrast it with the neoclassical growth model, in figure 8.1.

Panel a of figure 8.1 shows the neoclassical approach. Inputs of labor, materials, and the services of physical and human capital are transformed by the economy's aggregate production function to produce economic performance, as measured by total national income. The form of the production function depends on the economy's structure and its technology, but these things are hidden in a black box, whose only manifestation is the amount of output from given amounts of inputs.

---

1. This chapter relies on much of the evidence first introduced in chapter 2. Specific references are not repeated where they have already been given in the earlier chapter.

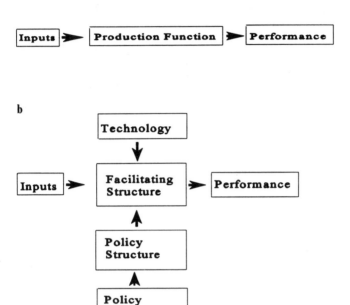

Figure 8.1
Comparison of neoclassical (a) and structuralist (b) growth models

Panel b shows our structuralist approach. Technology, the blueprints for the products we make, and processes by which they can be made are embodied in the structure, including the internal organization of the firm, the geographical location and concentration of industry, the infrastructure, and the financial system in order to produce economic performance measured by such variables as total national income, its distribution, and the total amount of employment and unemployment. Inputs are transformed by the facilitating structure to produce economic performance. Public policies are embodied in the policy structure, which can cause changes in the facilitating structure, technology, and inputs.

Technology
We define technology as consisting of

- *product technologies*, the specifications of the products that can be produced—where products refer to both intermediate and final goods and services;

- *process technologies*, the specifications of the processes that are, or currently could be, employed to produce these goods and services.[2]

Our definition of technology separates it from its embodiment in such things as capital goods and the organization of production which are part of the facilitating structure which we define below. The stock of existing technological knowledge (both applied and fundamental) resides in firms, universities, government research laboratories and other similar production and research institutions.

Facilitating Structure

The elements of the facilitating structure are

- all physical capital,
- people who embody human capital,
- organization of production facilities, including such things as the layout of factories and labor practices,
- managerial and financial organization of firms,
- geographical location of industries,
- industrial concentration,
- all infrastructure,
- private financial institutions and financial instruments.

The facilitating structure is the embodiment of technological and policy knowledge. Our concept of technology includes both those embodied in capital goods and those embodied in organizations (broadly defined).

Public Policy

This covers the specification of the objectives of public policy as expressed in legislation, rules, regulations, procedures, and precedents, as well as the specification of the means of achieving them as expressed in the design and command structure of public sector institutions from the policy force to government to international bodies.

Policy Structure

This covers the means of achieving public policies. These means are embodied in public sector institutions which are part of the policy structure. (Note the parallel with technology and its embodiment in capital

---

2. Both product and process technologies include those directed at producing new technological knowledge.

goods which are part of the facilitating structure.) It also includes the human capital related to the design and operation of public sector institutions (institutional competence).

Inputs and Economic Performance
Primary inputs of labor and raw materials are fed through the facilitating structure to produce the system's economic performance. It includes

- aggregate GDP, its growth rate, its breakdown among sectors, and among such broadly defined groupings as goods production and service production;
- GNP and its distribution among size and functional classes;
- total employment and unemployment and its distribution among such subgroups as sectors and skill classes.

Changes in technology typically have no effect on performance until they are embodied in the facilitating structure. Changes in any important technology, and specifically in GPTs, often induce pervasive changes in this structure. The full effects on performance will not be felt until all the elements of the structure have adjusted.

8.2 Effects on Technologies, Structure, And Policy

The majority of existing models of GPTs have assumed that their short- and long-run effects are determined by the GPT itself and its predetermined need for supporting technologies. While this may be a reasonable place to begin, we argue below that the effects are actually determined not only by the new GPT's characteristics but also by how it interacts with other existing technologies, the facilitating structure, and public policy.

8.2.1 Interactions among Technologies

Effects Arising from the GPT's Own Characteristics
Although the details differ across classes of GPTs, there are certain characteristic interactions shared by all GPTs, many of which are a direct consequence of the technological characteristics that we have already identified.

First, we can be sure that the technology will start in a relatively crude form with a limited number of uses (often one), and will only slowly develop the whole range of characteristics that we associate with a GPT. This initial evolutionary path is related to how agents learn about new technological ideas under conditions of uncertainty—by doing, by using, and by conscious experimentation. Second, as the GPT becomes more sophisticated and more generally applicable in terms of both range and variety of use, it will develop many Hicksian and technological complementarities. These characteristics have a number of implications, most of which are illustrations of the general observation that the GPT's impact on facilitating structure, policy, and macro performance will be slight at first and build up over time.

Effects That Depend on Interactions with Existing Technologies
In assessing the impact of a new GPT as it relates to existing technologies, the set of relevant questions includes: What is the productivity of the new GPT relative to the technologies it is displacing? Is the new GPT in competition with or complementary to each existing technology? *The answers will depend on the characteristics of both the new GPT and the existing technologies, some of which will be other GPTs.* Each set of relations between new and established GPTs will generate its own transitional effects. Some of the more important of these effects are discussed below.

Even when a technology that eventually becomes a GPT is introduced to meet some specific crisis, as was steam to deal with water in ever-deepening mines, it sooner or later begins to compete with technologies that are not themselves in crisis (e.g., sailing ships). Many of these older technologies are overcome by the new technology after a period of intense competition in which both technologies become more productive. As the new technology clearly demonstrates its superiority over the old, R&D money shifts away from the old technology so that its productivity stagnates. This shift of R&D is clearly not a cause of the introduction of the new technology but a consequence.

In some cases the new technology will quickly become superior to the older competing technology. For example, when electric motors challenged steam in factories, they quickly established their supremacy as a power delivery system once the unit drive was established. Few new steam-driven factories were build thereafter (although existing steam-driven factories lasted for decades). In other cases the initial margin of advantage is small and the transition correspondingly slower, as was the case when steam competed with water power in factories.

A new GPT often cooperates with an established technology that it eventually challenges. This was the case when early steam engines were sometimes used to lift water to help drive waterwheels, and early aeroplanes delivered passengers to the ports used by transoceanic liners.

When the new technology becomes dominant, it sometimes internalizes, or cooperates with, a revised version of the old technology. For example, the steam turbine, a combination of waterwheel and steam technologies, remains in use today as a electricity-generating device.

In some cases there may be little or no competition with the established technologies because the new technology fills a new niche. Regular, long-distance, transoceanic trade was created by the three masted sailing ship that had no established marine technology with which to compete over long distances. The internal combustion engine provided services that existing steam technologies could not—fast starting engines that were efficient at small horse power ratings and had relatively low weight/power ratios. Only when it was combined with electricity, did the hybrid diesel-electric engine seriously challenge steam on such major uses as ships and railways.

Sometimes a new GPT is complementary with a different type of existing technology. This was true, for example, of power and materials during the industrial revolution. Stronger materials were required before high-pressure steam engines could be perfected. It is also true of electricity and ICTs today.

In every displacement that we have considered, the full effect on productivity depends on the difference between the productivity of the new GPT and that of the technology it displaces. Since both technologies evolve, the effect is time dependant and cannot be predicted solely from a knowledge of the new GPT's own characteristics.

8.2.2 Interaction with the Facilitating Structure

We are concerned here with those elements of the facilitating structure that agents change in attempts to maximize the profit potential of a new GPT. At the outset, we note that major changes in the facilitating structure and in economic performance can be caused by changes in product and process technologies that are not themselves GPTs. For example, large structural changes followed the introduction of the telegraph. The new technology enabled the global coordination of goods and information leading to tighter international market integration. It also had profound effects on the railroad, domestic commerce, and other communications activities.

Channels of Impact
Although, all GPTs end up inducing changes in the economy's facilitating structure, GPTs in different classes typically make their effects felt through different channels.

ICTs. ICTs rarely have their first impacts on existing capital goods outside of the ICT industry itself. Much of the initial impact of new ICT technologies is typically on institutions, and methods of coordination at all levels of the economy. These changes often require, or enable, subsequent changes in product and process technologies, which in turn are embodied in new capital goods. Good examples of this dynamic were the drastic increases in tax revenues allowed by the introduction of writing in Sumer and printing in the Netherlands. The resulting alteration in the optimal unit of governance that followed on the introduction of writing ultimately had big effects on such things as agricultural production techniques, irrigation, and architectural technologies in Sumer. Similarly the invention of printing had important ramifications in the governance of the six northern provinces of the Netherlands. This led to new technologies particularly in the areas of finance and commerce.

Transportation. Changes in transportation technologies can often be employed without major alterations in the ways existing technologies are currently embodied in capital goods, or the ways machines are laid out on the shop floor. What does change almost immediately is the location of industry and those elements of the economy that deal with trade. In some cases, however, a new transportation technology may create entirely new industries or ways of doing business. For example, Chandler (1990) details the huge effects created by railroad technologies in the early part of this century.

Materials. Changes in materials often make their effects felt through quite different channels. They typically require, or facilitate, the direct redesign of many product and process technologies. Eventually the effects spread through the entire economy, inducing changes in institutions and methods of organization. For example, the invention of bronze had its initial effects on product technologies but eventually led to a major extension of the boundaries of the state, to a greatly increased importance of markets, and to a restructuring of the government (with power passing from the priesthood to lay rulers).

Power delivery systems. A new pervasive power technology usually has dramatic impacts. The use of new power sources often requires the redesign of physical elements of the facilitating structure including specific capital goods embodying task-specific machinery, the layout and location of the factory, and many elements of the economy's infrastructure including public institutions. Thus power delivery systems exert their effects through channels used by both materials and ICTs. For example, the introduction of the waterwheel induced many industries to employ new capital goods, required firms to locate beside a source of flowing water, and caused a drastic increase in the demand for river resources. The latter led to the development of complex systems of property rights over water and the institutions to support them.

Organizational technologies. Organizational technologies typically enter the facilitating structure as changes in the layout of the factory floor, or changes in management procedures. For example, although the introduction of lean production induced critical innovations in machine tools, it largely affected such things as the role of assembly line workers and the relationship between assembly plants and parts suppliers. After they are introduced, they typically induce a number of technological innovations in other categories, since the new organizational technology highlights new opportunities to adapt the facilitating structure.

Long-Term Changes in Structure

We now look at some examples that illustrate the major types of long-term changes induced in the facilitating structure by the most pervasive of new GPTs.

New capital. We have seen that the evolution of a new GPT is accompanied by important changes in a wide range of technologies. Changes occur both in the internal makeup of stand-alone capital goods and in technology systems. Often the latter requires restructuring of the cooperating capital goods before the full potential of the technology system can be realized. Many examples have been given earlier.

Human capital. Required skills change with virtually every GPT. In British mines the mule and horse handlers required much more skill than the drivers of the power-driven coal carriers that replaced them. Steam-driven factories and machines required skilled machinists, but the electric motor was more reliable, so repair and maintenance devolved to a few specialists. The modern computer requires skills that a significant part of the labor force does not have (although, as computers become more "user friendly," the skills required for their basic operation diminish steadily).

Reorganization of production facilities. A reorganization of the production process is usually required before the full potential of a new GPT can be realized. Early British textile factories were powered by people, animals, or water. The introduction of steam engines initially caused only a small increase in productivity. Large gains were not realized until the entire factory had been fully redesigned and its layout adapted to the power requirements of steam. Similarly it is taking decades for the potential of the computer to be realized on the factory floor through alterations in the whole structure of production. The same kind of reconfigurations are taking place in service industries where whole new divisions of labour are evolving around the new ICTs.

Reorganization of management practices. New technologies often require new management structures. For example, before computers, a typical firm had many middle-level managers who collected information and controlled its use. When computers were first introduced into firms, many senior managers complained that the new machines were lowering productivity. Only after years of experience were production and management procedures redesigned to take advantage of the power of computers. Many middle managers, whose jobs were to process information and pass it up and down the largely pyramidical management hierarchy, lost their jobs. As time passed, electronic data archiving, retrieval, and manipulation all began to take place on the computer, removing the need for elaborate paper filing systems. Also everything from product design to marketing has been altered by the use of the new ICTs. In the process many firms have been redesigned into more laterally linked units (Zuboff 1994).

Geographical relocation. New technologies often alter optimal locational patterns. Water power restricted production to sites with reliable, fast-moving water flows. Although steam factories could be located anywhere, the cost of transporting coal was of major importance, and the factory had to be located where the power was generated. With electricity, the generation and use of power was separated, freeing industry from the considerations of local power sources. Thus the efficient geographical location of production facilities changed radically when steam replaced water, and when electricity replaced steam.

Industrial concentration. Different technologies have different inherent scale economies that exert a major influence on industrial concentration. The use of water power required relatively small production units because the total horse power that any one site could generate was limited. Steam created large-scale economies, since the efficient size for a steam engine produced much more horse power than had the most efficient waterwheel. The computer induced the decentralization of many economic activities that were formerly done within one firm, reducing industrial concentration in many industries.

Changes in infrastructure. New technologies typically require new infrastructure. For example, some of the most important cost components of early steam engine installations were the sheds that housed the machine and coal, the foundational slab for the machine, and other microinfrastructure requirements. Economists are more familiar with the macroinfrastructure required by the steam engine, which included railway tracks, suitably designed ports, roads, and factories.

Transitional Impacts on the Facilitating Structure
We have seen that a new GPT induces major long-term changes in the facilitating structure. The structural adjustments associated with the introduction and evolution of a new GPT can be a long and conflict-ridden process. First, the required structural changes are subject to the same uncertainties as are the evolution of the GPT itself. The decisions that firms must take concerning the timing of its investment in new technology, geographic relocations, size of firm (and hence industrial concentration), and internal financial structure depend on many other structural adjustments. The uncertainty associated with the future evolutionary path of any new GPT also implies that there may not be an optimal response in the facilitating (or the policy) structures at each point in time. Agents in both the private and public sector may disagree about appropriate responses, even when all have the same information. Current debates about the restructuring of firms, labor practices, education systems, and government policies with respect to science, technology, industry, and regulation, all reveal genuine differences in assessment about the best responses to new technologies—differences not solely motivated by the desire to protect special interests.

Second, many elements of the facilitating structure are characterized by large sunk costs that delay adjustment. This was one of the main reasons for the long delay in electrifying the whole manufacturing sector in North America and Europe. This is also true of human capital. Many older workers find it difficult to retrain for the new technologies, and the full adjustment may have to wait until the older generation is replaced by a new generation trained in the requirements of the new technologies. In the process there will be uncertainty about the training that is appropriate.

Third, the required changes in the facilitating structure destroy many existing sources of rents and create many new ones. Those with vested interests in old sources of economic rent resist the changes. They often have substantial political power and long periods of conflict often occur. For example, many unions are currently resisting the reorganization of work induced by the lean production methods of Toyotaism. Similarly the Luddites, who were skilled craft manufacturers, resisted the integration and mechanization of production that accompanied the industrial revolution in England.

Fourth, there may be elements of structure that do not respond optimally because they are determined by noneconomic forces such as religion or public policy. The effects of a new technology on performance may be influenced by these unresponsive elements of the structure. For example, a public policy that resists the adjustment of the facilitating structure will result in different economic performance than would occur if those adjustments were left wholly to the market.

8.2.3 Interactions with Public Policy and the Policy Structure

Changes in Public Policy

Technological changes and the resulting change in the facilitating structure often require major changes in policy. Property rights need to be defined over new technologies, as well as over their cooperating factors—for example, streams to power waterwheels and airwaves for radio, television, and cellular phones. New policies are required with respect to labor practices, competition, and natural monopolies (whose identities change as technologies change). Changes in the volume and nature of foreign trade and investment call for new forms of international cooperation and control.

A detailed treatment of these issues is beyond the scope of this chapter. We provide three illustrations of the profound changes in existing public policies that are occurring in response to current technological changes. First, the current ICT revolution, and the globalization that it had facilitated, has induced alterations in many government policies. Sophisticated communications (and the vast amounts of short-term capital now in existence) make it impossible for governments to control international capital movements in the ways that they routinely did in the era of fixed exchange rates. Second, the increasing difficulty that governments face in dictating what their citizens will see and hear has curtailed the efficacy of information-restriction policies exercised in the interest of many varied purposes, from supporting a repressive dictatorship at one extreme to encouraging local cultural industries at the other. Third, the assets that confer many of today's national competitive advantages tend to be both created and highly mobile. This severely restricts any individual government's ability to adopt policies that affect these assets and that differ markedly from policies followed by other governments. Although this takes little space to say, its effects in limiting policy independence are profound.

Changes in Policy Structure
Changes are often needed in the institutions that give effect to public policy. This is particularly apparent today at the international level. ICT-assisted globalization is requiring international supervision of many issues involving trade and investment. The importance to most countries of a relatively free flow of international trade and investment has led them to transfer power over trade restrictions to supra-national bodies such as the World Trade Organization, the EU, the NAFTA, and a host of other trade liberalizing institutions. The interrelation of trade and investment brought about by the ICT revolution has caused modern trade liberalising agreements to be expanded to include measures to ensure the free flow and "national treatment" of foreign investment.

Today's typical government bureaucracy still has the hierarchical from of functionally defined departments that characterized firms in the Fordist era. Many of today's new governmental concerns typically cut across old boundaries. To function well in the future, governments need to be restructured to mimic the flatter organizations of firms. Governments typically find it more difficult to make the required adjustments than do private sector agents. (For further discussion, see Lipsey 1997.)

8.3 Effects on Performance

We now look directly at the long-run and transitional effects of new GPTs on the performance of the economy. We look briefly at labour markets and then concentrate on national income, although of course other performance variables are affected as well.

8.3.1 *Labor Markets*

In the long run, what happens to income inequalities will depend to a significant extent on the distribution of the human capital requirements in the new technology compared with the old. Assume, for example, that new technologies require a larger variance in the distribution of human capital than did old technologies—lots of low-skill and high-skill jobs rather than lots of medium-skill jobs. The inequalities in the distribution of income may then widen, even after all market adjustments have been made. Sophisticated measurement will show that this is a return to a different distribution of human capital, but the overall results will nonetheless be more inequality in the perceived distribution of income.

The transitional effects can be serious even if markets clear continually. What matters is the relative speeds of the demand- and supply-side changes. *If* the supply-side adjustments lag the demand-side shifts, those who adjust quickest to the new demands will find their skills in short supply, while those who adjust slowly, or not at all, will find their skills plentiful. In such a case the first group will enjoy increase in their relative wage rates. As a result the distribution of income may become more unequal. (To the extent that the market does not clear, the transitional effect will be higher unemployment rather than more income inequalities.) Once the pattern of demand has stabilized, the supply adjustments will catch up, and the inequalities in the distribution of income will narrow (and unemployment decrease).

This discussion suggests that we should be cautious in interpreting theories that build the same specific labor market effects into all new GPTs (as do most existing theories that deal with the labour market). What happens in the transition, and in the longer term, depends on how the pattern of labor demand that is associated with the new GPT relates to the pattern associated with the technologies it challenges or displaces. There is no reason why this relation should be the same, GPT by GPT. In other words, transitional and permanent labor market effects are more likely to be determined by the interrelations between a new GPT and the technologies it challenges than by some structure inherent in each individual GPT considered in isolation.[3]

8.3.2 Gross Domestic Product

The effects of GPTs on total output are of interest to economists for at least two reasons: their theorized effects in rejuvenating the long-term growth process (largely argued by economic historians), and their transitional effects, particularly showdowns (largely argued by theorists).

---

3. In chapter 11 of this volume Murphy, Riddell, and Romer use labor market observations to provide an estimates of the rate of technological change that are alternatives to the usual TFP calculations. To do this, they make the strong assumption that technological change always increases the ratio of the marginal product of high-skilled workers to that of low-skilled workers, which they measure by their relative wages. Finding no structural break in the trend of this relative wage, they conclude that there has been no acceleration in the rate of technological change.

Long-Term Effects

Economists have long debated whether or not economic growth can be sustained over the long term.[4] We distinguish between the effects of the development of a single GPT and the replacement of one by another. Consider a single GPT first. A GPT's development trajectory includes enabling a series of individual technologies. Investigators have fitted logistic curves to the experiences of many such individual technologies. Some have conjectured that the whole GPT will also follow such a curve, including the key point of eventually having an ever-slowing rate of productivity development. This conjecture would be correct, for example, if the individual technologies that each GPT could enable were strictly limited in number and each followed its own logistic curve. The overall GPT would then finally encounter falling productivity. If this were true for every GPT currently in existence, then growth would eventually slow and finally stop. Whether or not these conditions are found in practice is an empirical question—there does not as yet seem to be any slowing in the number of new technologies that electricity is enabling.

Now consider radical shifts from one GPT to another. When this happens, there is nothing to suggest any particular pattern in the effects on aggregate output or any other performance variable. This is because there is nothing in the economics of ideas, or in the physical and engineering properties of technologies, to suggest diminishing returns as technological knowledge accumulates generally, or, by implication, as one GPT replaces another. There is no reason to believe, for example, that there were larger effects on performance when steam replaced the waterwheel than when electricity replaced steam. There are no known limits to economic growth based on technological change, since new GPTs "rejuvenate" the growth process by providing a new set of evolving opportunities for a series of related technological advances. Further, as long as investment in capital goods continuously embodies new technological knowledge, there are no known limits to endogenously generated growth based on capital accumulation.[5]

Transitional Effects

Economic showdowns are the transitional effects that have most concerned theorists. It is often alleged that one sort of slowdown or another

---

4. This discussion is separate from the sustainability issue as defined by the Bruntland commission (1987).
5. The argument in the text is spelled out in much more detail in Lipsey and Bekar (1994) and is put into the context of historical controversies about the limits to growth in Lipsey (1994).

will occur whenever a new GPT replaces an existing one. Four obvious candidates for slowdowns are: a slowdown in the rate of GDP growth, a reduction in measured total factor productivity, a slowdown in the rate of increase in output per person hour, and a slowdown in real wage growth. Since each of these can behave differently, explaining any one does not necessarily imply explaining the others.

The main problem with any hypothesis concerning transitional effects is to establish a link between the evolutionary paths of individual GPTs and the observed macro behavior of the economy. Most existing theories sidestep this problem by assuming that only one GPT is in use at any one point in time and that its evolution determines the economy's macro behavior. In practice, however, several GPTs are operating at any one point in time—usually at least one in each of our classes—as well as many other important technologies. Each of these has its own development trajectory. The economy's macro behavior is influenced by the characteristics of all of the technologies currently embodied in the facilitating structure, most of which will be older technologies, at least until the new GPT gains supremacy. It is quite possible for all individual technologies in the economy, including GPTs, to follow a logistic pattern of productivity growth, while macro growth follows random variations around a more or less stable trend.

To get a macro experience that mirrored the experience of individual GPTs, one of three possibilities needs to be realized. First, the occurrence and development of the individual GPTs could just happen to be synchronized by chance—as appears to have happened with the current ICT and materials revolutions. Second, their paths may be synchronized by some coordinating mechanism. This is true in TEP theory, where individual technologies form part of a paradigm that itself behaves as if it were a single GPT. Third, there may be one dominant GPT whose influence is so pervasive that its behavior dominates the macro experience—the four most obvious candidates for this role over the past two centuries being the factory system, steam, electricity, and the modern ICT revolution.

In order to consider other issues, we assume away this important problem for the moment. We follow most existing theories in assuming that only one GPT exists at a time. How then might the introduction of a new GPT in this one-GPT world produce a slowdown in macro productivity growth? Two types of explanation have been advanced.

Type 1: Endogenous with respect to the GPT that it replaces. In this type of explanation, an existing GPT approaches the limit of its potential development, causing a slowdown in the rate of innovation and productivity growth. Either way, the "crisis" in the old GPT creates either the incentive for the development of a wholly new GPT or the opportunity for an existing technology that is at the early stages of its development to become a GPT. The new GPT has a slow start for all the reasons mentioned earlier. Then, as structural changes progress, innovation and productivity growth accelerate. In this theory, R&D shifts from the old GPT to the new when its development is reaching some inherent limit. The macro slowdown is the combined result of the slow growth associated with the exhaustion of the old GPT, and the slow development of a new GPT. The resulting growth path is generated by grafting together a series of logistic curves. Each dominant GPT spawns a successor only after its full development.

Such theories are suspect as universal explanations of slowdowns, since, as we saw in chapter 2, new GPTs do not typically arise our of general crisis associated with an existing GPT. First, some technologies that evolve into GPTs arise because of a very localized crisis in an old very specific technology, such as the pumping of water out of mines. In these cases much time passes before the evolving technology challenges many other existing technologies. Some of these challenges occur with technologies that are clearly not in crisis, as when the iron steamship challenged the wooden sailing vessel. Others occur with technologies that are approaching some physical limit, as when motor vehicles challenged horse-drawn vehicles as a method of transport—there was literally no more room for additional horsepower to be crammed into the already overloaded city streets, and horse manure was causing a significant health hazard. Second, other technologies that evolve into GPTs emerge from scientific research programs that are unrelated to the technologies they will challenge (e.g., electricity). Third, yet other technologies that evolve into GPTs are initially developed for noneconomic motives (e.g., the computer). Thus there are many reasons for the introduction of a new GPT, most of which have nothing whatsoever to do with an exhaustion of the potential of the current GPTs.

Type 2: Exogenous with respect to existing technologies. In this type of explanation, the early development of the new GPT can be either exogenous or endogenous to the economic system. The important point, however, is that the incentives are not created by any "crisis" in the current GPT. As far as users of the established technology are concerned, the new GPT arrives exogenously. As R&D switches from the old to the new GPT, the productivity growth of the old GPT slows and finally halts—but this is a result, not the cause, of the introduction of the new GPT. The new GPT's slow early development then produces a slowdown in measured productivity growth that is not related to the performance of the GPT that is being displaced.

A major problem with this approach is to explain why agents abandon the old technology in favor of the new when the old technology still has development potential and the net result of switching to the new is a current slowdown in productivity growth. This problem emerges in its starkest form if we use a set of strong assumptions. First, we assume that the resources currently devoted to creating future output are valued in the GDP at their discounted present value. Second, we assume that all output is produced under perfect competition so that marginal cost equals price everywhere. Third, we assume that there are no externalities. Fourth, we assume that the evolutionary path of each new technology is fully foreseen from the time it first comes into existence.

In these circumstances agents will only do R&D on, and/or put in use, a new technology if it adds to the present value of the expected stream of output. Assume, as in chapter 3, that a new technology is being researched but not yet used in production. If the immediate payoff to this R&D is low at the outset (the new technology is at the early flat stage of its logistic curve), firms will only persist in developing the new technology if they foresee the longer-term payoff that will occur when the GPT's evolution reaches the steep portion of its logistic curve. Given perfect foresight, and if the present investment in R&D is correctly measured, there should be no slowdown in the rate of growth of either GDP or TFP. Slowdowns of various sorts can be generated by departing from one or more of these four assumptions. We consider them in turn.

Helpman and Trajtenberg, and Howitt writing in chapters 3 and 9 in this volume, and many other economists who were not directly addressing GPTs, have argued that mismeasurement can produce observed slowdowns of various sorts. Three of the many possible sources of mismeasurement seem particularly important to us. First, by undertaking R&D on a new GPT, agents are investing in future output. Properly measured, this would show up as increased value of their firms today and hence in profits. Nonetheless, their actual measurement in the national accounts, as flows of current R&D *costs*, could produce a spurious slowdown. Second, the new GPT may be associated with an accelerated rate of depreciation of existing physical and human capital. Crude rules for calculating obsolescence will overestimate the true capital stock and lower measured TFP. Third, a new GPT is always associated with many new products, which tend to be imperfectly measured, particularly in the early stages of their development. This is likely to be serious when, as with the modern ICTs, so many of the new products and activities are in the service sector which tends be less adequately measured than the goods producing sectors. In all of these cases the slowdown is spurious. An important implication is that the main bottleneck to a better understanding of slowdowns may not be the development of more elaborate theories but more empirical research into the source, magnitude, and time patterns of mismeasurement.

A second class of explanations abandons the assumption of perfect competition and looks to an R&D-induced increase in the missallocation of resources to produce a slowdown. There are many ways in which this might occur. For example, Helpman and Trajtenberg assume that in the R&D sector marginal cost equals price, while some other sector is monopolistically competitive with marginal cost less than price. The introduction of a new GPT draws resources into the R&D sector, some of which come from the monopolistically competitive sector, increasing the missallocation of resources. This explanation requires that R&D behave competitively, even though much of it is done by oligopolistic firms, and that there be a net increase in R&D of sufficient magnitude to produce the observed slowdown.

A third class of explanations relies on externalities. It is possible, for example, for a new technology to be introduced despite conferring a net social loss, as long as the agents responsible for introducing the technological change are net beneficiaries. They rationally introduce the technology despite the net social losses which might partly show up as various forms of macro slowdowns. Furthermore the costs of structural change tends to be up front while the benefits are enjoyed by some in the present and everyone in future generations. Thus, although a cost benefit analysis showed that some proposed new technology should not be introduced because it conferred a net social loss, a second cost benefit analysis, made after it is in place, might show that the technology should not be removed. This allows for a sequence of GPTs, each one of which has a negative net present social value and causes an initial slowdown, but all are nonetheless accepted later in their evolutionary path as socially valuable.

A fourth class of explanations invokes uncertainty. All theories of GPTs in this book, and any others of which we are aware, assume perfect foresight. As we have argued at some length in both our chapters, the evidence is otherwise. Major technological changes are replete with uncertainties.

For example, faced with a choice between developing two alternative versions of some generic technology, some rational agents will back the first while others will back the second. Given this kind of uncertainty, it is possible that rational agents will choose a path of technological change that imposes a productivity slowdown—a change that they would not have made if they had known its full consequences.

An actual example drawn from Rosenberg (1982) can be used to show how such a mechanism might work (although in this particular case it would not be quantitatively important enough to show up at the macro level). In the early 1890s wooden sailing vessels were finally being replaced on all long-distance freight haulage routes. Several competing propulsion technologies were being rapidly improved. Shipowners faced a difficult decision, taken under much uncertainty, about whether to adopt the new technology now (at heavy capital expense) or wait until an improved or new version became available in the future. First, too early an adoption could leave owners saddled with technologies that became obsolete within a very short time. The subsequent early write-off of much capital investment could produce a slowdown because the expected value of output from the new capital did not materialize. Second, ship owners could elect to wait until technological advance slowed somewhat, and the relative advantages of the various new propulsion units were clearer. This behavior slows the rate of investment that embodies the new technologies in the facilitating structure causing a productivity slowdown. The uncertainty about adopting new technologies causes a rapid rate of technological progress to be associated with a slowdown in realized productivity growth.

The Evidence
What does the evidence from the technological and historical literature suggest about the issue of GPTs and productivity slowdowns?

First, we are sceptical of the methods used in many existing theories to generate slowdowns. Helpman and Trajtenberg's model of slowdown relies on swings in the allocation of labor between final production and R&D (which of course may stand for all types of learning). This requires three conditions for which we do not believe there is strong evidence: The introduction of a new GPT must be accompanied by a temporary growth in R&D which pulls significant quantities of resources from the goods production sector; on average, price must be closer to marginal cost in R&D than it is in all other activities; the reallocation into R&D and the differential between price and marginal cost must, in combination, be quantitatively large enough to produce an observed slowdown in the rate of growth of total output.

Second, as we have argued earlier, the impact of a new GPT on its own sector, let alone on the whole economy, depends on how it relates to the technologies currently in place. To know how a GPT will affect the growth rate, we need to know (1) its current and projected productivity relative to that of the technology it is replacing, (2) whether the new GPT is a substitute or a complement to extant technologies in other categories, (3) how the new GPT relates to the existing facilitating structure of the economy, and (4) the properties inherent in the GPT itself. By building the creation of slowdowns into the nature of the GPT itself (and its directly supporting technologies) most theories concentrate exclusively on condition 4.

Third, we have identified many GPTs, while, as far as we know, only three slowdowns have been alleged to have occurred over the last two centuries. The first is in the last part of the eighteenth century during the early stages of the factory system; the second is the in last part of the nineteenth century during the introduction of electricity; the third is in the period starting in the late 1970s alleged to be associated with the modern ICT revolution. The steam engine, the railroad, the motor car, the modern materials revolution, and the laser have not been suggested as possible slowdown culprits. It seems clear therefore that all GPTs cannot be associated with macro-productivity slowdowns. This poses the challenge of identifying those that have the potential to cause slowdowns, a challenge that we discuss further in the next section.

One thing we can be sure about is that a new potential GPT will rarely have immediate, significant, productivity effects until complementary technologies have been developed and structural adjustments made. In a technologically dynamic society we would always expect to see some degree of mismatch between the existing facilitating structure and the current technology. Ceteris paribus, the more rapidly technology is changing, the greater the degree of mismatch. But many steps are needed to link these structural adjustments to productivity slowdowns. This is not to suggest that GPTs *never* cause slowdowns. We argue only that GPTs do not always cause slowdowns. Further, when a GPT is the source of an observable macroeconomic slowdown, we would expect to find certain characteristic patterns within the development trajectory of the GPT itself, and interactions with other technologies, the existing facilitating structure, and policy institutions.

8.4 Further Challenges

8.4.1 Two Approaches to Model Building

The ways in which the assumed characteristics and consequences of GPTs are accommodated in theories reveals two approaches to model building. We illustrate these with the conjecture that GPTs cause productivity slowdowns.

In the first approach, one starts with some observed or assumed result and then builds a model that will produce that result. The value of this approach is partly to act as a feasibility check. If economists cannot build a model based on some observed characteristics of GPTs that produce a productivity slowdown, we must be sceptical of the notion that GPTs inevitably cause slowdowns. This feasibility test is passed by any model that produces the result from assumptions that capture some of the characteristics believed to be associated with GPTs. The test is, however, a weak one, since an ingenious modeler can usually produce some sort of a model that will generate almost any desired result if given enough latitude in the permissible assumptions. Such a model will have much more value if it predicts other events or associations that it was not constructed to explain.

In the second approach, one builds a model with the characteristics of GPTs and sees what follows. The value of this approach is that it will almost certainly make predictions that the model was not constructed to produce. Under this approach, slowdowns are a possible outcome of the exercise, not the objective. If models that contain an increasing number of characteristics of GPTs stubbornly refuse to predict slowdowns, this will increase skepticism about the proposition that any real technology will produce observable slowdowns.

8.4.2 Challenges for Theory

The first generation of models designed to capture GPTs and their effects are typically quite sophisticated theoretical constructs. This should not surprise us. As we have stressed, GPTs are complex phenomena, both with respect to their own properties and in their relationships to other technologies, the facilitating structure, and public policy. To have captured any of the significant properties of GPTs in a viable model is an important accomplishment. Nonetheless, these models tend to omit some of what appear to be key elements of GPTs. The relation between existing models and the evidence coming from both empirical and historical sources pose many challenges for future theorizing.

Hierarchies

Some investigators have suggested a hierarchy of GPTs. Two technologies can be placed in a technological hierarchy if one is necessary for the other, but not vice versa. For example, the automobile and the airplane, at least in the forms that we know them, would not have been possible without the internal combustion engine, while that engine would have existed without either of the other two GPTs. Bresnahan and Trajtenberg model a single technology tree with one GPT sitting at the top. This is a vast improvement over theories in which technology is flat. But, as our empirical cases have earlier shown, the relationship is even more complex than a single technology tree. First, a GPT, while sitting at the top of some technology tree, is also horizontally related to other GPTs. Second, it also may be vertically related to another GPT.

Channels of Effect

We have seen that although GPTs in different categories usually all end up changing many existing technologies as well as most elements of the facilitating structure, different classes of GPTs usually make their effects felt through different channels. This is not modeled in any current theory, since they all deal with only a single type of GPT. The result is that they miss the different channels through which different GPTs affect other technologies, the various elements of the facilitating structure, and economic performance.

Technological Complementarities

Our study of the technology literature suggests that complex complementarities, both Hicksian and technological, are an important characteristic of GPTs. In the existing theories, however, most complementarities and transitional effects are modeled as price and productivity changes in extant GPTs. We have argued earlier that a price change does not capture the physical alterations that the introduction of a new technology causes in the facilitating structure. Thus such explanations leave out important information. Even in this limited domain, relations of substitutability rather than complementarity are more often the ones that are modeled. Eaton and Lipsey (1989) have shown the limitation inherent in using the Dixit-Stiglitz utility function in formal models of monopolistic competition. When this function is used to model production, as in Helpman and Trajtenberg, the most important implication is that all the ancillary technologies that support the GPT are substitutes for each other. We argued earlier, however, that the various types of ancillary technologies are more commonly complementary to each other, although competing versions of each of these ancillary technologies will be substitutes. Modeling this mixture of competitiveness and complementarity is an important challenge.

Bresnahan and Trajtenberg (1995) use strategic complementarities to model technological ones. It is possible to view their second derivative (strategic complementarity) as capturing all of the effects that we have detailed, since the model is specified at such a high level of abstraction. As such, this type of model misses much of the detail that is useful in explaining the effects that GPTs have on the performance of an economy. Bresnahan and Trajtenberg were not, however, trying to capture the details of different types of complementarity. Rather they were modeling a coordination problem, for which the generic complementarity is sufficient.

A major theoretical challenge is to devise a way of dealing with the fractal-like, interlocking nature of technology, which leads to complementary effects when changes occur in some element of any technology system. Each new technology tends to have a time dependent set of complementarities and competitive relations with many other technologies. For each pair of technologies, these relations depend on whether they stand in horizontal or vertical relationship to each other, whether they compete with or complement each other, and whether the impact of the GPT on the other technology is Hicksian or technological in the senses defined earlier.

Competition among Technologies

In most theories there is only one active GPT and one arriving GPT, with a well-defined transitional period in which the old GPT is replaced by the new. Such transitional dynamics are often modeled as being identical over all GPTs, being repeated as each new GPT is introduced. We think an interesting challenge is presented by trying to capture the effect of heterogeneous GPTs competing with existing heterogeneous technologies (including some GPTs). As soon as the new GPT is introduced in most theoretical models, the old GPT is no longer the target of any R&D activity. In some cases it is immediately pulled out of final goods production. While these sorts of stylisations may be necessary to generate stable time paths, they are not what we see in history. In many cases in which a new GPT is introduced, several distinct old technologies "fight back" and undergo sudden bursts of development.

Facilitating Structure
Our concept of the facilitating structure outlines the linkages between changes in technology and changes in performance. At this stage of its development our model does not detail the specific links between technology and each element of the facilitating structure. Some of the links run directly from technology to elements of the structure, while others run between the various elements of the structure. Another important set of linkages runs from changes in structure back to changes in technology. Foremost on our agenda is to specify these linkages more tightly so that the complex paths by which changes in technology ultimately affect economic performance may be more fully understood.

Origins
We should be wary of theories in which a single firm, or even a single sector, is modeled as a GPT "supplier." First, the early evolution of GPTs is typified by such pervasive uncertainties that it is impossible for agents to ascertain whether or not any single new technology (not to mention a collection of related technologies) will become a full-blown GPT, a restricted-purpose technology of some importance, or a technological midget. Second, as we have seen earlier, GPTs evolve as a series of related technologies, often in different sectors doing very different jobs. Third, the strong complementarities that surround a GPT make it exceedingly difficult for a single firm acting alone to capture the benefits and incur the costs of developing such a complex technology. Last, the evidence shows that a GPT has never been completely controlled beyond its initial stage by a single firm, and never has one firm been responsible for the creation of more then one GPT.

To date, most theories have chosen to make GPTs either exclusively endogenous or exclusively exogenous. Both assumptions conflict with the facts that GPTs have many different sources, both economic and noneconomic. For some purposes, particularly when one is concerned mainly with the consequences of GPTs, treating them as wholly exogenous is an acceptable simplification. However, the rich experience of the origins of GPTs poses a challenge to those who wish to include in their theories an explanation of the generation of GPTs.

Uncertainty
Much turns on how the information sets of agents are modeled. Most existing theories ignore risk, let alone uncertainty. They assume that agents recognize a GPT the moment it is introduced and have full infor-

mation about its future development. As we have already observed, GPTs are typified at the beginning of their development by uncertainties with respect to the evolution of such variables as range of use, productivity, and costs. As the GPT develops, and supporting innovations are made, the future becomes somewhat more predictable. Developers of some of the GPT's subtechnologies will come to know that the technology will have a significant impact. The details of how the GPT will develop, however, may still be uncertain. Once the GPT becomes an important technology, much of the uncertainty about its role will have been removed. Even then, however, no one will have any clear idea of the nature of the new GPT that sooner or later will challenge it, of when that challenge will occur, and of how the existing GPT will fare in the competition that will ensue.

While it would be no small task, a major step forward would be to find ways to include uncertainty into models of GPT dynamics. We are aware of the staggering difficulties associated with modelling uncertainty in any but the simplest environments, and in anything but the most abstract manner, but we also recognize the high payoff that might be gained if some operational version of uncertainty could be incorporated in models of GPTs.

8.4.3 Outlook

Issues concerning technological change are in the forefront of attempts to explain long-term growth. Only lately, with the growing interest in GPTs, have attempts been made to develop formal theories of growth that take account of the structures of technology systems. Whether one talks of GPTs, or just pervasive technologies, structural issues are at the heart of an understanding of growth as driven by technological change. Given the complexity of these structural relations, it seems fruitful to push research on two fronts. The first is formal theories that initially will have to employ crude abstractions from the complex reality. The second is appreciative theories that can incorporate more complexity but lack the explicit rigour of more formal theories. Both approaches can shed light on the phenomena we discuss in this volume; both can assist the other by providing things to explain and by checking on explanations provided by the other approach.

Developing satisfactory theories of GPTs is not a task that will be completed quickly or easily. It seems to us that the theoretical research program should be to extend existing models, and/or to develop new models, to capture more of what we know empirically about GPTs rather than elaborating and generalizing just because we are able to do so. In this program there would be a large payoff to the development of new models that are designed to capture more of the characteristics of GPTs in their assumptions, and then explore the implications of those assumptions.

References

Chandler. 1990. *Scale and Scope*. Cambridge, MA: Belknap Press.

Eaton, B. C., and R. G. Lipsey. 1996. Beyond neoclassical competitive economics. In *The Foundations of Monopolistic Competition and Economic Geography: Selected Writings of B.C. Eaton and R.G. Lipsey*. Cheltenham: Elgar.

Eaton, B. C., and R. G. Lipsey. 1989. Product differentiation. In R. Schmallensee and R. Willig, eds., *A Handbook of Industrial Organization*, vol. 1. Amsterdam: North Holland.

Lipsey, R. G. 1994. Markets, technology change and economic growth, *The Pakistan Development Review* 33: 327–52.

Lipsey, R. G. 1997. Globalization and national government policies: An economists view. In J. Dunning, ed., *Globalization, Governments and Competitiveness*.

Lipsey, R. G., and C. Bekar. 1995. A structuralist view of technical change and economic growth. In *Bell Canada Papers on Economic and Public Policy*, vol. 3. Proceedings of the Bell Canada Conference at Queen's University. Kingston: John Deutsch Institute.

Rosenberg, N. 1982. *Inside the Black Box: Technology and Economics*. Cambridge: Cambridge University Press.

Woomack, J. P., D. J. Jones, and D. Roos. 1990. *The Machine that Changed the World*. New York: Rawson Associates.

World Commission on Environment and Development. 1987. *Our Common Future*. Oxford: Oxford University Press. Chair was Norwegian Prime Minister Gro Harlem Brundtland.

Zuboff, S. 1984. *In the Age of the Smart Machine: The Future of Work and Power*. New York: Basic Books.

9   Measurement, Obsolescence, and General Purpose Technologies

Peter Howitt

9.1   Introduction

The introduction of a new general purpose technology often brings with it an accelerated pace of technological change, as a result of the dynamic complementarities and scope for improvement that Lipsey, Bekar, and Carlaw (chapter 2) have identified as two of the salient characteristics of a GPT. Accelerated technological change can cause major dislocations and adjustments, at least at the level of individual households and business firms. For example, costly adjustment to rapid change in information technologies became a commonplace of everyday experience in the 1980s and 1990s. David (1990) has argued that a similar pattern was also characteristic of electrification in the early part of the twentieth century.

This chapter examines the question of whether the cost of adjusting to the acceleration of technological change brought on by a new GPT might have an effect at the macroeconomic level as well as at the individual level. In particular, it examines channels through which faster technological change might cause a slowdown in measured economic activity.

Others have investigated possible *structural* links between GPTs and economic stagnation. Lipsey and Bekar (1995) argue, for example, that the benefits of a new GPT will not be realized on a significantly widespread level until the supporting structure of the economy has adapted to it. Freeman and Perez (1988) go further with their concept of a technoeconomic paradigm shift. Helpman and Trajtenberg (1994) argue that the benefits of a new GPT will not be realized until enough secondary innovations have occurred to make it more profitable to use than earlier GPTs.

This chapter was inspired by a comment made by Luc Soete during my presentation of on an earlier paper (Howitt 1996) in a 1995 conference, in which he insisted that more attention should be paid to the problem of capital-obsolescence. Karen Park provided helpful research assistance, and an anonymous referee provided many useful remarks on an earlier draft.

The purpose of this chapter is to step back from the structural details of GPTs and to investigate the effects that can be produced by a new GPT simply because the introduction raises the pace of technological change. More specifically, there are two channels through which faster technological change can cause an empirically significant slowdown in measured output. The first is the result of a measurement problem that I identified in an earlier paper (Howitt 1996) as the knowledge-investment problem. That is, the output of knowledge-creating activities such as R&D are for the most part excluded from the national accounts. Thus, when the opportunities opened up by a new GPT induce people to reallocate resources away from producing goods, which are included in the accounts, toward producing knowledge, which isn't, there may be a spurious fall in measured GDP growth.

The second channel is capital-obsolescence, which could cause a *non*-spurious fall in GDP growth. Growth in output per person depends on capital accumulation as well as knowledge accumulation. A faster pace of capital-embodied technological change will tend to reduce the net rate of capital accumulation, by causing both physical and human capital to become obsolete more rapidly. I argue below that the overall effect of this combination of faster knowledge accumulation and slower capital accumulation is likely to be a fall in GDP growth that lasts for many years.

Since technological change and obsolescence are central concepts of Schumpeter's process of creative destruction, the model I construct below in order to develop these arguments is an extension of the model of growth through creative destruction presented in Aghion and Howitt (1992, 1996). This extension introduces out-of-steady-state dynamics in order to focus on the timing of the effects of introducing a new GPT. It also introduces population growth in order to calibrate the model numerically to the U.S. economy. The model eliminates the counterfactual positive scale effect of a higher population on growth, which was present in earlier R&D-based growth models,[1] by assuming, as in Young (1995), that the enhanced profitability of innovation in a more populous economy will be dissipated by a larger number of industries over which innovations must be spread.

With the use of this model, the chapter assesses the relative importance of the two effects identified above—the spurious slowdown in measured GDP growth caused by the failure to measure knowledge investment, and the actual slowdown caused by accelerated capital-obsolescence. The

---

1. See Jones (1995).

chapter concludes that the actual slowdown is likely to be the stronger of the two effects. In the baseline simulation a new GPT that raises the productivity of R&D by 50 percent, for a period long enough to shift the economy's aggregate production function by 100 percent, will cause *actual* GDP to fall below its no-shock path for 28 years, with a maximal percentage shortfall of 6.6 percent. The knowledge-investment problem exaggerates the shortfall, but only by a maximum of 2.5 percentage points.

Moreover, in measuring the effect of a new GPT on *national income* growth, the knowledge-investment problem will probably be more than offset by another measurement problem related to obsolescence, namely that the depreciation rates used in the national accounts are unlikely to be adjusted quickly to reflect the accelerated capital-obsolescence induced by faster technological change. On balance, taking into account both of these measurement problems, measured national income is likely to *understate* the slowdown caused by the new GPT.

Section 9.2 discusses various national income accounting problems raised by innovation and technological change, with special focus on the knowledge-investment problem. Section 9.3 explains in simple terms how the obsolescence created by more rapid innovations can have a short-run effect of reducing the growth rate of GDP. Section 9.4 constructs the theoretical growth model. Section 9.5 calibrates a numerical version of the model to the U.S. economy and presents the simulation results. Section 9.6 contains some concluding remarks.

9.2 Knowledge and the National Income Accounts

To measure economic growth at the level of an entire country, economists have little alternative to using the aggregate measures of national income and output generated by the system of national accounts. But there is a fundamental problem with using the national accounts for measuring growth as opposed to cycles. That is, whereas economic growth, in per capita terms, is attributable to the accumulation of knowledge as well as the accumulation of capital goods, the national accounts are designed to measure only goods. The theoretical foundation of national income accounting supposes that knowledge is fixed and common, and that all that need be measured are prices and quantities of commodities. Likewise we do not have any generally accepted empirical measures of such key theoretical concepts as the stock of technological knowledge, human capital, the resource cost of knowledge acquisition, the rate of innovation, the rate of obsolescence of old knowledge, and the like. Because of our inability to measure properly the inputs and outputs to the creation and use of knowledge, standard measures of GDP and national income can give a misleading picture, especially during a period of transition when a new GPT has enhanced the opportunities for knowledge creation.

In particular, there are at least three major national income accounting problems concerning the (non)measurement of knowledge that bias our measures of GDP growth.[2] The first is what I call the *knowledge-investment problem*. That is, the output of knowledge resulting from knowledge-investment activities, such as formal and informal R&D, is typically not measured at all because it does not result in an immediate commodity with a market price. From the Haig-Simons point of view, the creation of knowledge ought to be treated like the creation of capital goods, since in either case there is an expenditure of resources that could alternatively have been used to produce current consumption but that has instead been devoted to the enhancement of future consumption opportunities. Yet the national accounts include no category of final expenditure that would capture a significant amount of the annual increment to society's stock of knowledge the way it captures the annual increment to society's stock of capital, except for the output of the educational sector, and for R&D undertaken by or sold to the government sector. None of the new knowledge generated by R&D undertaken by business firms on their own account, which includes most of industrial R&D, results in a direct positive contribution to current GDP or to the current value added of that sector of the economy, as would happen if the resources had instead been devoted to the creation of new capital goods.

To make this point more explicitly, consider the case of a firm that hires additional R&D workers at a cost of one million dollars during the current year, the only result of which is a new patent received at the very end of the year, which will enable the firm to earn additional profits in future years, whose expected discounted value is two million dollars. Since firms are not permitted to capitalize R&D expenditures, this sequence of events will not result in any increase in output from that sector as far as the national accounts are concerned. Likewise, from the income side of the accounts, although there has been an additional one million in wages and salaries (assuming that the workers were hired from out of the labor force) there has been an exactly offsetting decrease in profits, since the expendi-

---

2. The following discussion summarizes the lengthier analysis of Howitt (1996) where I also discuss a fourth measurement problem, which I called the "knowledge-input" problem, whose primary effect is on measures of productivity growth rather than GDP growth.

ture by the firm resulted in no increased current revenue. If, instead of the patent, the workers had produced a machine worth two million dollars, GDP would have been higher by two million dollars.[3]

Of course, to the extent that R&D results in more or better goods being produced, it does eventually affect measured GDP. But new knowledge should also be counted as output when it is created, just as physical investment is counted when it is created even though it eventually has a further effect on GDP by increasing the potential to produce other goods.

Furthermore, to the extent that R&D results in better goods, many of its future effects on GDP will not be measured, because of a second major measurement problem, the *quality-improvement problem*. As many writers have observed, when knowledge creation within business firms results in improved goods and services, the practical difficulties of dealing with new goods and quality improvements in constructing price indexes imply that much of the resulting benefit goes unmeasured.

The third problem is the *obsolescence problem*. If standard measures of GDP ought to include a separate investment account for the production of knowledge, then by the same token, NNP or national income ought to include a deduction corresponding to the depreciation of the stock of knowledge that takes place as it is superseded, or otherwise reduced in value, by new discoveries and innovations. Furthermore the creation of new knowledge is also a factor accounting for the depreciation of existing physical capital. Depreciation is a notoriously difficult concept to account for in any case. The timing and extent of replacement investment are endogenous variables that the national income accountant can only capture in rough measure by applying simple mechanical formulas. But the problem becomes even more acute when a wave of innovations accelerates the rate of obsolescence of old knowledge and capital. If measured depreciation rates are changed at all, it will only be with a long lag.

If an economy were in a steady state, the most serious of these measurement problems would be the quality-improvement problem. Much of the growth of productivity and output in the long run is the result of product innovations that generate new and improved goods whose contributions to output are only partially measured. Gordon (1990), for example, has estimated that correcting properly for quality improvements

---

3. The working group that produced the International *System of National Accounts 1993* considered recommending the capitalization of R&D expenditures for just these reasons. In the end they decided to drop the recommendation, in view of the problems of measuring and evaluating such an intangible investment, although they did recommend the setting up of satellite R&D capital accounts on an experimental basis.

in capital goods alone would at least double the growth rate of aggregate real investment in the United States over the period 1947 to 1983. Many of the gains from better capital goods do eventually get reflected in GDP growth when the improved capital goods boost output in other sectors. But even then the problem will distort measured productivity growth in different sectors, as when the airline industry is credited with productivity growth that actually occurred in the airframe and engine industries. Furthermore, to the extent that the improved capital goods allow other sectors to create new and improved products, the productivity gains may not even be measured in those sectors because of the quality-improvement problem, as when more powerful computers allow banks to produce a better quality of services.[4]

During a period of adjustment to a new GPT, however, the problem most likely to have a significant impact on measured GDP growth is the knowledge-investment problem. Because the opportunities for knowledge-creating activities are enhanced by the arrival of a GPT, workers spend less time producing goods and more time producing knowledge. The fall in goods production is recorded in the national accounts, but the rise in knowledge production is not. This can result in a spurious fall in measured GDP growth.

When computers first started to change the way work is done throughout the economy, for example, there was a long period of learning that had to be undertaken. At first people looked for ways in which the new tool could simply replace old ones without a radical change in operating techniques. Although some gains were obtainable in that direction, the added cost of information-service departments were often larger than the benefits. Gradually, through a process of trial and error, people are now learning to exploit the enormous potential of computers, but for many years there were no visible productivity gains associated with the adoption of sophisticated information technologies.

The time people have spent learning to use computers efficiently, and all of the associated costs of training and experimentation, constitute unmeasured knowledge investments. If they had been measured, there is no doubt that GDP growth rates would have been higher over the past two decades than they were. Moreover at least some of the slowdown in output growth that has occurred over this period must be attributable to the fact that this unmeasured knowledge investment has come at the cost of reduced goods production, which has been measured. Indeed, from a broader perspective, the costly restructuring of firms, and the sectoral

---

4. This point is made forcefully by Griliches (1994).

reallocations involved in learning better to exploit new fundamental technologies, also constitute unmeasured knowledge investments, with similar spurious effects on measured output growth.

Even beyond this, however, the reason why more workers go into knowledge-creating activities when a new GPT is introduced is that the return to these activities rises relative to the return to goods production. Thus it seems likely that the knowledge investment that goes unmeasured is even greater than the fall in goods output that is measured. So it seems likely that the effect of a new GPT could be to raise the growth rate of actual output at the same time as it reduces the growth rate of measured output.

It is possible that the quality-improvement problem could also cause a slowdown of measured GDP growth during the introduction of a new GPT. Just as it causes part of economic growth to go unmeasured in a steady state, so it will cause much of the surge of economic growth resulting from better computers and related goods to go unmeasured. Some of this problem has been dealt with by the adoption in the United States of hedonic measures of quality improvement in computers. But similar measures have not been undertaken in such sectors as the electronic equipment industry that manufactures chips.

Griliches (1994) claims moreover that the fruits of the information revolution have been used disproportionately in sectors where quality improvements are next to impossible to measure. He estimates that over three-quarters of the output of the computer industry is used in what he calls the unmeasurable sectors. Furthermore the information revolution is contributing to an increase in the relative size of the unmeasurable sectors, which Griliches estimates now accounts for 70 percent of GDP in the United States.

However, Baily and Gordon (1988) have argued that the quality-improvement problem cannot account for a lot of the slowdown in productivity that took place in the early 1970s, mainly because the failure properly to measure quality improvements has been too steady over time. Whether Griliches or Baily and Gordon are closer to the truth is an open question on which I have nothing more to say here.

The obsolescence problem is even less likely to produce a measurement bias in GDP growth because no deduction is made for depreciation in computing GDP. However, it can moderate or even overturn the effect of the knowledge-investment problem on measured growth in NNP or national income. Indeed the net increase in society's stocks of capital and knowledge resulting from the information revolution would be overstated if the accelerated obsolescence of preexisting capital and knowledge were not taken into account. For example, the development of the personal computer greatly reduced the value of mainframe computers, secretarial skills, typewriters, batch-programming knowledge, and many other items of human and physical capital. Moreover the fact that new versions of each producer's computers are now appearing every six months reflects an acceleration in the rate at which old computer equipment gets scrapped. Thus, if we were to solve the knowledge-investment problem without dealing with the obsolescence problem, we would certainly overstate the gain in NNP and national income taking place during a technological transition, even though measures of GDP would not be affected by the omission.

9.3 Obsolescence

Obsolescence induced by a new GPT can produce a slowdown in economic activity that is not spurious, by reducing the rate of new capital accumulation, a key determinant, especially in the short run, of economic growth. To see how this works, consider the familiar neoclassical growth model of Solow and Swan, with Harrod-neutral (purely labor-augmenting) technological progress, in which the stock k of capital per efficiency unit of labor grows according to the equation

$$\frac{dk}{dt} = sk^\alpha - (\delta + \beta + n + g)k, \tag{1}$$

where s is the saving rate, k^α is the production function determining output per efficiency unit, with $0 < \alpha < 1$, δ is the rate of physical capital depreciation, β the rate of capital-obsolescence, n the rate of population growth, and g the (exogenous) rate of technological progress. The first term on the right-hand side of (1) is gross capital accumulation per efficiency unit, and the second term includes the reduction the stock of capital through depreciation and obsolescence as well as the increase in the number of efficiency units.

The growth rate G of real output per person is the growth rate of output per efficiency unit (i.e., of k^α) plus the rate of technological progress:

$$G = \alpha \frac{1}{k}\frac{dk}{dt} + g = \alpha s k^{\alpha-1} - \alpha(\delta + \beta + n + g) + g.$$

In the long run, the capital intensity k will approach a constant steady-state value, output per efficiency unit will therefore stop growing, and the growth rate G will equal the rate of technological progress g. When technological progress goes up therefore, the long-run effect is to raise the rate of economic growth point for point. Even in the short run, *if the other parameters of the model are independent of the rate of technological progress*, faster technological progress will raise growth, although with an effect that is less than point for point. Specifically, the impact effect will be

$$\frac{\partial G}{\partial g} = 1 - \alpha > 0.$$

In practice, however, the other parameters of the model will not remain unchanged when the rate of technological progress increases. In particular, the rate of obsolescence β will surely increase, especially if the more rapid progress comes about from the introduction of a new GPT that displaces old production methods that use old capital equipment. Thus there will be an additional negative impact effect on dk/dt over and above that of the more rapid growth in efficiency units. Because of this the overall impact effect is likely to be a reduction in the rate of economic growth.

Suppose, for example, that β is proportional to g. Then from the above equation for G we have the impact effect

$$\frac{\partial G}{\partial g} = 1 - \alpha\left(1 + \frac{\beta}{g}\right).$$

The effect is more likely to be negative the larger is the parameter α, which measures the share of capital in GDP, and the larger is the ratio β/g of obsolescence to growth.

Suppose that we interpret capital as including human as well as physical capital. (Human capital is also subject to obsolescence during periods of rapid technological change, through the process of deskilling.) Then the reasoning of Mankiw (1995) makes $\frac{2}{3}$ a conservative estimate of α. For β, take the value that Caballero and Jaffe (1993) estimated as the annual rate of decline in the value of a US firm that doesn't innovate, namely 0.036. For g, take the average growth rate of output per worker between 1948 and 1991, namely 0.02. Thus we arrive at the estimate $\partial G/\partial g = -0.87$. That is, a rise in the rate of technological progress from 2 percent per annum to 3 percent would have the impact effect of reducing the growth rate of output per person from 2 percent per annum to 1.13 percent.

The critical parameter in this calculation is β, the rate of capital-obsolescence. Caballero and Jaffe's estimate of 0.036 was based on an analysis of the NBER R&D master file, a database covering 567 large U.S. firms over the period 1965 to 1981. They classified the firms into 21 different technology sectors, based on the nature of the firms' patents, and within each sector estimated an equation predicting the change in a firm's value relative to the total value of the sector, as a function of the flow of new patents from the firm and the flow from other firms in the same sector. Based on this estimated equation, Caballero and Jaffe calculated by how much a firm's value would have decreased per year on average if it had not produced any patents. This measure of the rate of obsolescence varied from 0 in several sectors to 25 percent in the drug sector, with an (unweighted) average across sectors of 3.6 percent.

The firms in this sample may be more R&D-intensive than the average U.S. firm, yet the estimate misses the obsolescence of firms in sectors where patents are not so important, like computer equipment. Moreover even less R&D-intensive firms face a threat of obsolescence if they do not continually modernize their capital and restructure their organizations when others are doing so.

Whether the same rate of obsolescence applies to human as to physical capital can perhaps be questioned. Earnings profiles (e.g., Katz and Murphy 1992, p. 49) suggest that people with more experience have greater earnings given their education level, but one would somehow have to control for on-the-job-learning investments before jumping to the conclusion that human-capital obsolescence is not a major problem. Clearly there are many examples (secretaries, middle management, mainframe computer programmers, electrical engineers) where the information revolution has forced people to undertake substantial knowledge investments to avoid economic losses. In the absence of data on such investments I have mechanically applied the Caballero-Jaffe estimate to human as well as physical capital.

In any event, a negative impact effect of productivity growth on output growth would still exist even if there were *no* obsolescence of human capital. Indeed, if the obsolescence rate were 0.036 on half of K and 0 on the other half, then β would be 0.018 and $\partial g/\partial G = -0.27$. In order to eliminate the negative impact effect on growth, the average obsolescence rate on all types of capital would have to be reduced to 1.0 percent per annum. Moreover, if, as seems likely, the marginal obsolescence effect of faster productivity growth resulting from a GPT is higher than the average, then the average depreciation rate on capital could be even less than 1.0 percent per annum for a negative impact effect still to exist.

While these numerical calculations are suggestive of an important effect, they leave a lot out of the picture. In particular, a rise in the productivity of the research technology and a prospective fall in the growth rate of output per person will likely feed back on the rate of innovation. Moreover it would be helpful to have some idea of how these effects play out over time. How long, for example, would one have to wait for the growth rate to return to its initial level? And how long would it take for real income per person to return to the level of the no-shock path? How would the answer to these questions be affected by taking into account that the enhancement of the research technology is not a permanent phenomenon but is part of the adjustment to a single new GPT? To address these questions, as well as to quantify the relative importance of the measurement issues discussed in the previous section, we need an empirically implementable non-steady-state endogenous growth model that includes innovation, capital accumulation and an endogenous rate of capital-obsolescence. This is the task of the following section.

9.4 An Endogenous Growth Model

This section lays out a non-steady-state version of a growth model that allows for population growth. In this model there is a single final output, produced by labor and a continuum of intermediate goods, according to the production function:

$$Y_t = Q_t^{\alpha-1}\left(\int_0^{Q_t} A_{it} x_{it}^\alpha \, di\right) N_t^{1-\alpha}, \qquad (2)$$

where Y_t is gross output at date t, Q_t is the measure of how many different intermediate products exist at t, $N_t = e^{nt}$ is labor input, assumed identical to the population that grows at the exogenous rate n, x_{it} is the flow output of intermediate product i, and A_{it} is a productivity parameter attached to the latest version of intermediate product i. The presence of the factor $Q_t^{\alpha-1}$ in (2) eliminates any productivity gain resulting from the proliferation of products.[5] Final output can be used interchangeably as a consumption or capital good or as an input into a research technology.

Growth in the number of products is assumed to take place as a result of serendipitous imitation. Each imitation produces a new product whose productivity parameter is identical to that of a randomly chosen existing product. Each person in the economy has the same exogenous propen-

---

5. As Benassy (1998) has recently pointed out, most endogenous growth models that use this Ethier-Dixit-Stiglitz framework assume arbitrarily that the return to product variety is determined by the same parameter as the elasticity of substitution α. I have chosen here to set the return to product variety equal to zero, independently of α. This extreme value simplifies the analysis without materially affecting the results.

sity to imitate. Hence the flow of imitation products can be written as $dQ_t/dt = \xi N_t$, $\xi > 0$. This implies asymptotic convergence of the number of products per person to the constant $q = \xi/n$. Hence I assume that $Q_t = qN_t$ for all t.

Each intermediate product is produced using capital, according to the production function

$$x_{it} = \frac{K_{it}}{A_{it}}, \tag{3}$$

where K_{it} is the input of capital in sector i. I divide K_{it} by A_{it} in (3) to indicate that successive vintages of the intermediate product are produced by increasingly capital-intensive techniques.

The local monopolist that produces each intermediate product is the innovator or imitator that created the latest generation of the product. The local monopolist in each sector produces with a cost function equal to $\omega_t K_{it} = \omega_t A_{it} x_{it}$, where ω_t is the cost of capital, and a price schedule given by the marginal product: $p_{it} = \alpha A_{it} x_{it}^{\alpha-1} q^{\alpha-1}$. Standard profit maximization yields the implication that all intermediate sectors will supply the same quantity, namely

$$x_{it} = x_t \equiv q^{-1}\left(\frac{\omega_t}{\alpha^2}\right)^{1/(\alpha-1)}, \tag{4}$$

and that each local monopolist will earn a flow of profits proportional to its productivity parameter A_{it}, namely

$$\pi_{it} = A_{it}\pi_t \equiv A_{it}\alpha(1-\alpha)q^{\alpha-1}x_t^{\alpha}. \tag{5}$$

Assume that the cost of capital always adjusts so as to maintain equality between the supply of capital K_t and the demand $\int_0^{Q_t} K_{it}\,di = x_t \int_0^{Q_t} A_{it}\,di = x_t Q_t A_t$, where A_t is the average productivity parameter across all intermediate products. Then the common output of all intermediate products will be

$$x_{it} = x_t = \frac{K_t}{A_t Q_t} = \frac{k_t}{q}, \tag{6}$$

where $k_t\ (\equiv K_t/A_t N_t)$ is the capital stock per efficiency unit of labor. (By definition each unit of labor produces A_t efficiency units.) Substituting from (6) into (2) yields the Cobb-Douglas aggregate production function:

$$Y_t = K_t^{\alpha}(A_t N_t)^{1-\alpha} = k_t^{\alpha} A_t N_t. \tag{7}$$

Improvements in the productivity parameters come through an innovation process that uses final output as the only hired input. The aggregate flow of innovations μ_t at date t is

$$\mu_t = \lambda \left(\frac{R_t}{A_t^{\max}}\right)^\gamma Q_t^{1-\gamma}, \qquad 0 < \lambda,\ 0 < \gamma < 1, \tag{8}$$

where R_t is the input of goods into research, $A_t^{\max}$ is the leading-edge technology parameter (the maximal value of the productivity parameters A_{it} in the economy), and λ is a parameter indicating the productivity of R&D. The parameter γ is assumed to be less than one to allow for the decreasing returns that several studies have found in R&D.[6] The appearance of the leading-edge parameter $A_t^{\max}$ in the right-hand side of (8) represents the force of increasing complexity; as technology advances, the resource cost of further advances increases.

The same input (adjusted for the leading-edge productivity parameter) will be used in research in each intermediate sector, namely $p_t \equiv R_t/Q_t A_t^{\max}$, because the prospective payoff is the same in each sector. Specifically, each innovation at date t results in a new generation of that intermediate sector's product which embodies the leading-edge productivity parameter $A_t^{\max}$. Thus the Poisson rate of innovation ϕ_t in each sector at date t will be

$$\phi_t = \frac{\mu_t}{Q_t} = \lambda p_t^\gamma. \tag{9}$$

The productivity-adjusted level of research will be determined by the arbitrage condition that the marginal cost of an extra unit of goods allocated to research equal the marginal expected benefit. The latter in turn is just the product of the value of an innovation V_t and the marginal product of research in producing a flow of innovations $\partial \mu_t/\partial R_t$. Hence we have the research arbitrage equation

$$1 = \lambda \gamma p_t^{\gamma-1} v_t = \gamma \frac{\phi_t}{p_t} v_t, \tag{10}$$

where v_t is the productivity-adjusted value of an innovation: $v_t = V_t/A_t^{\max}$.

I assume that with each innovation, the innovator buys an initial stock of capital as determined by the above optimality conditions and enters into Bertrand competition with the previous incumbent in that sector,

---

6. For example, Arroyo, Dinopoulos, and Donald (1994) and Kortum (1993).

who by definition produces an inferior quality of good. Rather than face a price war with a superior rival, the incumbent sells its capital for scrap value and exits. Having exited, the former incumbent cannot threaten to re-enter. This is why the local monopolist can always charge the unconstrained monopolist price without worrying about competition from earlier vintages of the product.

Accordingly the value of an innovation at date t is the expected present value of all the future profits to be earned by the incumbent before being replaced by the next innovator in that product. Using (5) and (6) we can write this as

$$V_t = \int_t^\infty e^{-\int_t^\tau (r_s + \phi_s)\,ds} A_t^{\max} \alpha(1-\alpha) q^{-1} k_\tau^\alpha \, d\tau, \tag{11}$$

where r_s is the instantaneous rate of interest at date s. The instantaneous discount rate applied in (11) is the rate of interest plus the rate of creative destruction ϕ_s; the latter is the instantaneous flow probability of being displaced by an innovation. Differentiation of (11) with respect to t yields the following differential equation in the productivity-adjusted value of an innovation:

$$\frac{dv_t}{dt} = -\alpha(1-\alpha) q^{-1} k_t^\alpha + (r_t + \phi_t) v_t. \tag{12}$$

Growth in the leading-edge parameter $A_t^{\max}$ occurs as a result of the knowledge spillovers produced by innovations. That is, at any moment of time the leading-edge technology is available to any successful innovator, and this publicly available (but not costless) knowledge grows at a rate proportional to the aggregate rate of innovations. The factor of proportionality, which is a measure of the marginal aggregate impact of each innovation on the stock of public knowledge, is assumed to equal $\sigma/Q_t > 0$. I divide by Q_t to reflect the fact that as the economy develops an increasing number of specialized products, an innovation of a given size with respect to any given product will have a smaller impact on the aggregate economy. Putting all this together yields a rate of technological progress equal to

$$g_t = \frac{1}{A_t^{\max}} \frac{dA_t^{\max}}{dt} = \phi_t \sigma = \lambda \rho_t^\gamma \sigma. \tag{13}$$

Because the distribution of productivity parameters among new imitation products at any date is identical to the distribution across existing products at that date, one can show that the ratio of the leading edge to the average productivity parameter will asymptotically approach the constant $1 + \sigma$.[7] Thus I assume that $A_t^{\max} = A_t(1 + \sigma)$ for all t, so the rate of growth of the average productivity parameter A_t will also be given by (13).

Rather than assume intertemporal utility maximization on the part of a representative consumer, as is usual in endogenous growth theory, I make the simpler, and in my judgment more empirically sensible, assumption of a fixed saving rate, as in the Solow-Swan model. Hence the rate of change of the capital intensity variable k_t is given by the above equation (1). To make this differential equation operational, however, we must discuss the determination of the rate of obsolescence β.

To this end I suppose that all capital is at least to some extent designed specifically for the sector in which it is used. That is, if it is dismantled and sold elsewhere, each unit, which was bought at a price of unity relative to consumption, is only worth the fraction η. In the case of human capital the interpretation of this assumption is that the fraction η is specific to the firm that hires the owner (or to that firm's technology). Hence the rate of obsolescence per unit of time per unit of capital is just the flow probability that the unit of capital will be scrapped ϕ_t times the loss in value $(1 - \eta)$ when scrapped. Putting this together with the technological progress equation (13) yields

$$\frac{dk}{dt} = s k_t^\alpha - [\delta + n + \phi_t(1 - \eta + \sigma)] k_t. \tag{14}$$

The cost of capital ω_t is the sum of three components—interest, depreciation, and obsolescence. Straightforward economic reasoning from the conceptual framework described above yields the formula

$$\omega_t = r_t + \delta + \phi_t(1 - \eta). \tag{15}$$

According to (4), (6), and (15),

---

7. By assumption, imitations do not affect the average A_t. Each innovation replaces a randomly chosen A_{it} with the leading edge $A_t^{\max}$. Since innovations occur at the rate $\mu_t/Q_t = \phi_t$ per product and the average change across innovating sectors is $A_t^{\max} - A_t$, we have

$$\frac{dA_t}{dt} = \phi_t(A_t^{\max} - A_t).$$

This and equation (13) imply that the ratio $\Omega_t = A_t^{\max}/A_t$ evolves according to

$$\frac{1}{\Omega_t}\frac{d\Omega_t}{dt} = \phi_t \sigma - \phi_t(\Omega_t - 1).$$

It follows that as long as ϕ_t is bounded above zero, Ω_t will converge asymptotically to $1 + \sigma$, as asserted in the text.

$$r_t + \phi_t = \omega_t - \delta + \eta\phi_t = \alpha^2 k_t^{\alpha-1} - \delta + \eta\phi_t. \tag{16}$$

According to (9) and (10),

$$\phi_t = \lambda^{1/(1-\gamma)}(\gamma v_t)^{\gamma/(1-\gamma)} \equiv \phi(v_t, \lambda), \quad \frac{\partial \phi}{\partial v_t} > 0, \quad \frac{\partial \phi}{\partial \lambda} > 0. \tag{17}$$

Equations (16) and (17) can be used to reduce the differential equations (12) and (14) to the two-dimensional system:

$$\begin{cases} \dfrac{dk_t}{dt} = sk_t^\alpha - [\delta + n + \phi(v_t, \lambda)(1 - \eta + \sigma)]k_t \\ \dfrac{dv_t}{dt} = -\alpha(1-\alpha)q^{-1}k_t^\alpha + [\alpha^2 k_t^{\alpha-1} - \delta + \eta\phi(v_t, \lambda)]v_t \end{cases}. \tag{18}$$

The system's rest point is a steady-state equilibrium with balanced growth in the usual sense. In this steady state the growth rate of output per capita is the rate of technological progress, which according to (13) and (17) can be expressed as an increasing function $\phi(v, \lambda)\sigma$ of the steady-state innovation-value v.

The system's phase diagram is illustrated in figure 9.1. The locus of steady-state capital stocks is downward sloping because an increase in the value of an innovation will raise the equilibrium rate of creative destruction and hence raise the rate of obsolescence, thus causing a fall in capital intensity for the same reason as in the Solow-Swan model. The locus of steady-state innovation values is upward sloping because an increase in capital intensity reduces the cost of capital that will be faced by a prospective innovator, as well as reducing the equilibrium rate of interest applied by researchers to discount future profits. The system exhibits the usual saddle-point properties.

9.5 Effects of More Rapid Technological Change

As I mentioned at the start of the chapter, I am not modeling any of the structural aspects of a GPT that others have identified as critical. Instead, I am focusing on one aspect of a new GPT, namely that its introduction and diffusion are associated with an increase in the rate of technological innovation. In the model just developed, a new GPT simply means an increase in the productivity of research, as represented by the parameter λ. The increase can be thought of as reflecting the dynamic externalities and scope for improvement that Lipsey, Bekar, and Carlaw (chapter 2) identify as two of the defining characteristics of a GPT. The new GPT

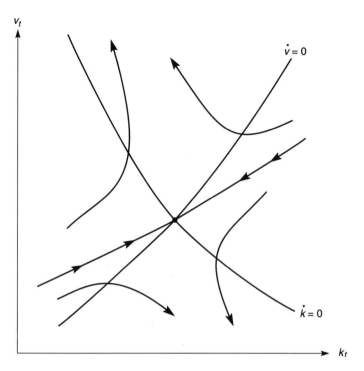

Figure 9.1
In the absence of any disturbance, capital stock (k_t) and the value of an innovation (v_t) approach their steady-state values monotonically

arrives initially in a crude form but opens up windows of opportunity for innovations aimed at realizing its potential. This indirect way of modeling a GPT also satisfies two of the other criteria of Lipsey, Bekar, and Carlaw, namely that the GPT is widely used (the productivity of R&D is assumed to rise in all sectors) and enters radically (in the numerical simulations below, the aggregate production function will eventually shift out by 100 percent as a result of the GPT).

Figure 9.2 illustrates the dynamic effects of a permanent increase in the productivity parameter λ of the research technology. Both steady-state loci shift down, but the steady-state capital-stock locus shifts further down than the steady-state innovation-value locus.[8] The increased rates of obsolescence and productivity growth stimulated by the rise in λ will cause k to start falling. The prospect of lower k in the future causes the value of an innovation v to fall discontinuously at first, and then the gradual realization of that prospect causes a subsequent further decline in

---

8. See appendix A.

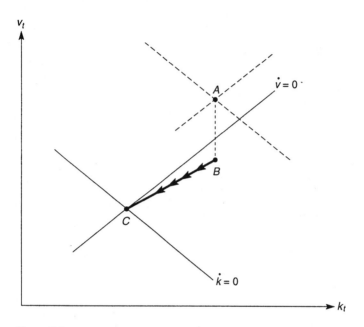

Figure 9.2
Reaction to an unexpected permanent increase in the productivity of R&D, starting from an initial state A

v. The new long-run equilibrium will have lower values of both k and v, but the overall effect on growth of the rise in λ and the fall in v will be to raise the steady-state rate of growth. Whether or not the growth rate goes down at first will depend, as in the Solow-Swan model, on the precise parameter configuration.

Figure 9.3 illustrates the effects of the more relevant conceptual experiment in which the research-technology parameter λ rises but only temporarily, reverting at some known future date to its original value. The system will follow a rebound trajectory lying above the new saddle-path CB, with the value of an innovation and the stock of capital both falling at first but then both rising after λ goes back down to its original value. The date at which the system reaches the saddle path of the original system must coincide with the date at which λ reverts to its original value. Some time before that date the value of an innovation will start to rise in anticipation of the rise in the stock of capital, which, as we have explained before, will raise the incentive to research. Thus, as is usual in rational expectations models, turning points in the forward-looking asset price will lead turning points in the historically predetermined stock variable.

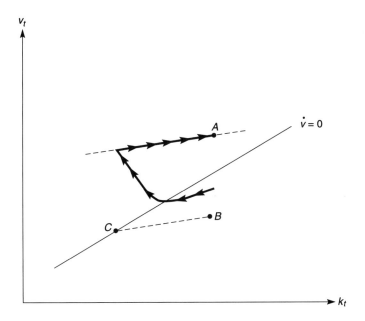

Figure 9.3
Reaction to an unexpected temporary increase in the productivity of R&D, starting from an initial steady state A

In the case illustrated in figure 9.3, if economic growth falls when the initial rise in λ occurs, it will rise when the movements in k and v are reversed. To the extent that the knowledge-investment problem leads to an underestimate of economic growth, as argued in section 9.2, the underestimate will be most pronounced near the end of the burst of innovations, that is, after the value of an innovation has started to rise in anticipation of the resumption of positive capital accumulation but before the productivity of research has fallen back to its original value. This is likely to be the time when the most resources are diverted from producing goods to producing knowledge, and hence the time when the failure to account for the production of knowledge in the national accounts is likely to show the greatest distortion.

While the dynamics of this model may seem complicated, the impact effect on growth of the increase in the technology parameter λ will be qualitatively the same as the impact affect $\partial G/\partial g$ of the simpler Solow-Swan model laid out in section 9.3. Equations (7) and (14) imply that as in the simpler model, the growth rate of GDP per person at any date will be

$$G = \alpha \frac{1}{k}\frac{dk}{dt} + g = \alpha s k^{\alpha-1} - \alpha(\delta + n + g + \beta) + g,$$

where in this case $g = \sigma\phi$ and $\beta = (1-\eta)\phi$.

The only way that a change in the innovation parameter λ impinges on this equation is through the innovation rate ϕ, which affects productivity growth g and the depreciation rate β in the same proportion. So the impact effect on growth is

$$\frac{\partial G}{\partial \lambda} = \frac{\partial g}{\partial \lambda}\left[1 - \alpha\left(1 + \frac{\beta}{g}\right)\right].$$

The term in square brackets is exactly the expression for the impact affect $\partial G/\partial g$ derived in section 9.3. Since $\partial g/\partial \lambda > 0$[9], then whether or not the new GPT causes an initial slump will depend on exactly the same factors as before. In particular, if $\alpha = \frac{2}{3}$ and the initial value of g is 0.02, then a slump will occur whenever the initial obsolescence rate β exceeds 1 percent per year.

To get an idea of how long the slump is likely to persist, the rest of this section produces a numerical version of the theoretical model of section 9.4, calibrated to the U.S. economy. As in section 9.3, I assume that $\alpha = \frac{2}{3}$ and that the economy is initially in a steady state with a growth rate $g_0 = 0.02$ and a rate of capital-obsolescence $\beta_0 = 0.036$. I assume a population growth rate $n = 0.01$. For notational simplicity I also assume an imitation rate $\xi = 0.01$[10] so that the number of products per person q equals unity. I also suppose that in the initial steady state the volume of resources used in research is 5 percent of GDP. This is twice the average for the U.S. economy over the period from 1959 to 1988, according to NSF data on total R&D. The knowledge-investment problem identified above suggests that the actual ratio is higher than measured but not by how much.

I set the depreciation rate $\delta = 0.036$. This number yields a sum of depreciation and obsolescence equal to 19.8 percent of GDP. Since I am interpreting capital to include physical and human capital, and since the share of capital α in GDP is approximately twice what one finds in the U.S. data, depreciation and obsolescence on physical capital alone would be a little under 10 percent of GDP, which is approximately what one finds in U.S. data as measured depreciation. Thus the depreciation figure is

---

9. This can be seen in figures 9.2 and 9.3. Indeed, if $\partial g/\partial \lambda \leq 0$, the impact affect would not be to make the capital intensity k start falling.

10. As I explain in my discussion of table 9.2, the behavior of GDP is independent of the assumed value of the imitation rate ξ, given all the other specified parameter values.

Table 9.1
Details of calibration

Specified values		
β_0	Initial obsolescence rate on capital	0.036
g_0	Initial growth rate of output per person	0.02
α	Capital's share in GDP	2/3
n	Population growth rate	0.01
ξ	Imitation rate	0.01
$(R\&D/Y)_0$	Initial R&D expenditure as a fraction of GDP	0.05
δ	Physical depreciation rate on capital	0.036
η	Scrap value as a fraction of replacement value	0.5
r_0	Initial rate of interest	0.09
Implied values		
ϕ_0	Initial exit rate of incumbent firms	.072
σ	Size of innovations	.278
s	Saving rate for total capital—human plus physical	0.28
k_0	Initial capital stock per efficiency unit	20.65
q	Number of intermediate products per person	1
y_0	Initial GDP per efficiency unit	7.527
$(\delta + \beta_0)k_0/y_0$	Measured depreciation as a fraction of GDP	.198
k_0/y_0	Initial capital-output ratio	2.743
v_0	Initial (productivity-adjusted) value of an innovation	10.325
γ	Elasticity of innovations with respect to R&D	.396
λ	Initial productivity parameter of R&D	.117

Note: See appendix B for the derivation of implied values.

in line with experience under the assumption that the depreciation rates used in the national accounts have reflected accurately the sum of depreciation and obsolescence on average.

I choose a ratio η of scrap value to replacement value equal to $\frac{1}{2}$. (As we will see in the discussion of table 9.2, the value of η does not play the critical role that one might expect.) I also suppose that in the original steady state the rate of interest is 9 percent, the average real rate of return on equity over long periods of time in the United States.

These specifications are enough to solve the model numerically. (See appendix B for details.) Table 9.1 gives an indication of the implied values of various magnitudes in the initial steady state. The value of the R&D elasticity γ is near the centre of the range typically estimated in the literature (e.g., Kortum 1993). The saving rate is about double what one finds in U.S. data, which is appropriate given that our comprehensive measure of capital is roughly twice that of the physical capital whose accumulation is measured in the savings data. The main problem appears to be the capital/output ratio, which is only about 10 percent higher than the measured ratio of physical capital to output. This can be corrected by allowing a more conventional rate of interest of 4 percent, with no significant change in the simulation results, but in order to match the depreciation data, I must then set the rate of physical depreciation equal to zero.

To simulate the effects of the introduction of a GPT, I perturb this initial steady state by increasing the research-technology parameter λ by 50 percent and leaving it at its higher value for as long as it takes for the average productivity parameter A to rise by 100 percent, which turns out to be 23 years. Figure 9.4 shows the resulting trajectory of the productivity-adjusted value of an innovation and the capital stock per efficiency unit. As indicated earlier, the capital stock falls for the full 23 years during which the new GPT is being adopted, while the value of an innovation stops falling after 15 years. After year 23, both variables rise monotonically to their original values. The fall in the capital stock per efficiency unit during the initial 23 years is 36 percent of its original value. Figure 9.5 shows that this 23-year period of introduction of the new GPT is a period of intense research and of an unusually high rate of obsolescence.

Figure 9.6 illustrates the resulting trajectories of economic growth (G) and technological progress (g). While technological progress shoots up with the introduction of the new GPT, the rate of economic growth falls and remains below its initial 2 percent value for 14 years. As figure 9.7 illustrates, the level of output per person remains lower than it would have been without the introduction of the GPT for 28 years, that is, 5 years longer than the period of rapid technological change. The maximum percentage shortfall below the no-shock path is 6.6 percent, and it occurs in year 15.

Table 9.2 shows how the size and duration of the slump depend on the parameters of the model. It shows the maximal percentage shortfall below the no-shock path, and the number of years before GDP catches up to the no-shock path, when each parameter is changed, one at a time, by 50 percent of its baseline value. As the analysis of section 9.3 indicated, the critical parameter values are the first three in the table. A larger initial depreciation rate, a smaller initial growth rate, or a larger capital-elasticity of output would each make the slump longer and more severe.

Given the value of these parameters, the results are not sensitive to the others. Even the scrap value ratio η is not particularly important; what matters instead is the rate of obsolescence $\beta_0 = (1 - \eta)\phi_0$. If a higher

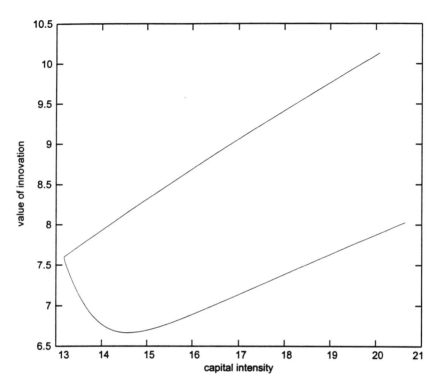

Figure 9.4
Computed path following a temporary 50 percent increase in the productivity of R&D

scrap value were assumed, with no change in the rate of obsolescence, then the implied initial innovation rate ϕ_0 would be higher, with little effect on the behavior of GDP. (See appendix B for details.) The imitation rate ξ has no effect at all on the behavior of GDP; instead, it just has an inversely proportional effect on the whole path of innovation-values v_t.

Although table 9.2 indicates that the size and duration of the slump are both sensitive to the first three parameters, it also indicates that a slump will occur even if our calibration is off by 50 percent in any one of the parameter values, with the exception of α, the share of capital in GDP. If α is reduced by 50 percent, then the initial impact effect of GDP is slightly positive. The critical value at which the slump disappears is $\alpha = 0.357$. However, as long as we take capital to include human capital, this critical value is far below any realistic estimate.

It should also be noted that if the fixed saving rate s were replaced by the Ramsey assumption of intertemporal utility maximization by a representative household with isoelastic utility of consumption and a fixed rate

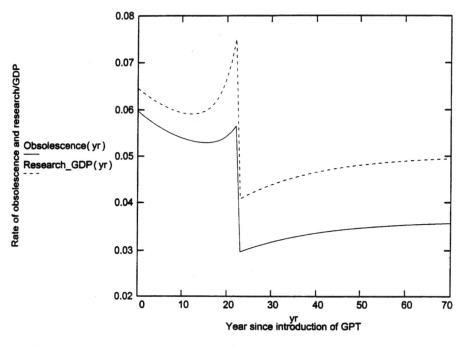

Figure 9.5
Computed effects on the rate of obsolesence and the R&D/GDP ratio

of time preference, then the slump could either be amplified or dampened. Indeed the wealth effect of anticipated income gains would cause people to save less, thus further reducing growth, but on the other hand, the rise in the rate of return to investing in R&D would tend to offset this wealth effect.

The simulation exercise also sheds light on some of the measurement problems discussed in section 9.2. Consider, for example, the knowledge-investment problem. In the theoretical model the measure of GDP that I have defined includes all of output, whether it is consumed, put into capital, or used as an input to research. Suppose, however, more realistically that none of the resources used in research were counted as final output because of a failure of the national accounts to measure the accumulation of knowledge. Then measured GDP per person would not be $k_t^\alpha A_t$, as assumed above, but $[k_t^\alpha - \rho_t(1+\sigma)q]A_t$. The percentage gap of measured output below its no-shock path would be significantly greater than if output were measured correctly, throughout the period during which the GPT was being adopted, because of the reallocation out of the measured goods sector and into the nonmeasured knowledge sector. Figure 9.8

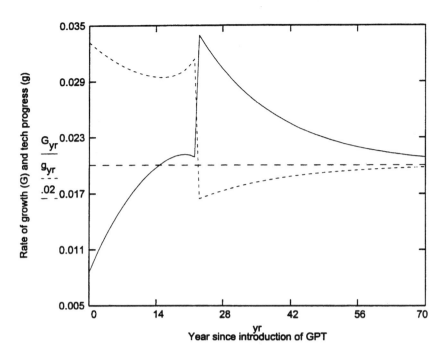

Figure 9.6
Computed effects on the rate of economic growth and the rate of technological progress

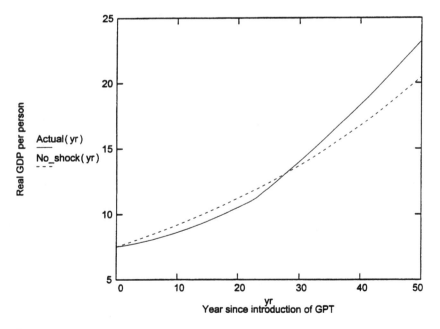

Figure 9.7
Computed effects on the path of per-capita GDP

Table 9.2
Sensitivity analysis on the duration and maximal size of the shortfall of actual GDP below the no-shock path

		Size (%) 6.6		Duration (years) 28	
Baseline		50% smaller	50% larger	50% smaller	50% larger
β_0	(0.036)[a]	1.4	11.2	16	32
g_0	(0.02)	12.4	3.4	56	18
α	(2/3)	0	Undefined	0	Undefined
α[b]	(2/3)	1.4	18.5	13	41
n	(0.01)	6.9	6.4	29	28
ξ	(0.01)	6.6	6.6	28	28
$(R\&D/Y)_0$	(0.05)	6.6	6.7	29	27
δ	(0.036)	7.7	5.8	31	26
η	(1/2)	6.9	6.3	28	29
r_0	(0.09)	6.5	6.9	29	27

a. The numbers in this column are the baseline values used in the main simulation described in the text.
b. Because the model makes no sense when α is increased by 50 percent, this row shows the effect of a 25 percent increase or decrease in α.

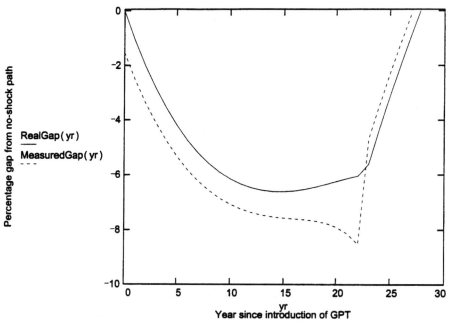

Figure 9.8
Computed effects on the actual and measured GDP gaps, where R&D investment is not measured in the actual figures

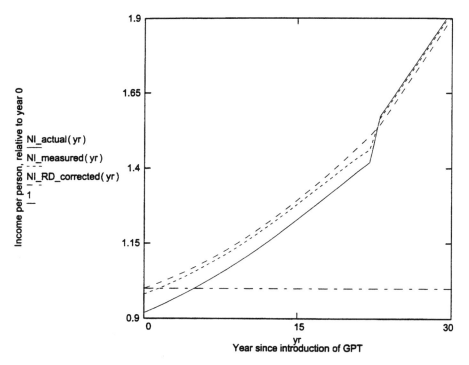

Figure 9.9
Computed effects on actual and measured per-capita national income

shows that the measured gap on average throughout this period is over 1 percentage point larger than the actual gap. In year 22 the measured gap reaches its maximum value of 8.5 percent, when the actual gap is only 6.0 percent.

The quality-improvement problem, as discussed above, may or may not significantly affect the pattern of adjustment to the GPT. All we know is that it distorts the long-run rate of growth. However, the obsolescence problem does create a significant distortion in the level of national income. In fact it more than offsets the measurement bias resulting from the knowledge-investment problem. Figure 9.9 shows the behavior of measured national income, actual national income, and R&D-corrected measured national income (i.e., national income corrected for the knowledge-investment problem but not for the obsolescence problem), all per person relative to their pre-shock values. Measured national income is computed on the assumption that depreciation is calculated at the same rate as before the shock and that the pre-shock measured depreciation rate is the correct rate: $\delta + \beta_0$.

The increase in obsolescence causes actual national income to fall below its initial value (not just below its no-shock reference growth path) for 5 years, the initial drop being 8.1 percent. Measured national income also falls, but only by 1.9 percent and only because of the knowledge-investment problem. In fact, as the path of R&D-corrected measured national income shows, correcting for the knowledge-investment problem alone would significantly improve the measure of GDP but worsen the measure of national income, since the failure to include knowledge investment compensates, at least partially, for the much larger measurement problem raised by obsolescence.

These simulation results suggest that although both of the main channels that we have identified are at work creating a slowdown in measured GDP growth, the illusory component of the slowdown (attributable to the knowledge-investment problem) is small relative to the real component (attributable to capital-obsolescence). Indeed, while the maximal real gap between GDP and its no-shock reference path is 6.6 percent, the maximal amount by which the gap would be overstated is only 2.5 percentage points (in year 22), even under the extreme assumption that no knowledge investment whatsoever is included in measured GDP.

Moreover the results suggest that failure to adjust the depreciation rates used in the national accounts to reflect the accelerated obsolescence caused by a new GPT is potentially a more serious problem than the failure to include any knowledge investment. Clearly, under the assumption that no adjustment is made at all to the depreciation rates used in the national accounts, the overall effect is for the slowdown in national income growth to be *under*stated by conventional measures. The maximal understatement is 6.2 percent, and it occurs in the first year when measured national income falls by only 1.9 percent even though actual national income falls by 8.1 percent. During the entire period of the slump, the average understatement is 4.2 percent of initial national income.

Table 9.3 shows that these conclusions regarding the size of the two measurement problems (the knowledge-investment problem and the obsolescence problem) are robust to variations in parameter values. In all except two cases the maximal real GDP gap (as reported in table 9.2) exceeds the maximal overstatement of the gap (as reported in table 9.3). The two exceptional cases were the extreme cases in which the real gap was only 1.4 percent. Table 9.3 also indicates that the problem of measuring obsolescence more than offsets the knowledge-investment problem in all cases, in the sense that the average gap between measured and actual national income during the period of the slump is in all cases positive.

Table 9.3
Sensitivity analysis on the size of measurement problems affecting the GDP gap and national income

		Maximal difference between measured and real GDP gap (%)		Average gap between measured and actual national income (% of initial level)	
		2.5		4.2	
Baseline		50% smaller	50% larger	50% smaller	50% larger
β_0	$(0.036)^a$	3.8^c	2.2	1.6	5.1
g_0	(0.02)	2.5	2.5	3.8	4.5
α^{**}	(2/3)	2.6^c	4.2	2.3	2.9
n	(0.01)	2.4	2.5	4.0	4.2
$(R\&D/Y)_0$	(0.05)	0.9	5.9	4.7	3.5
δ	(0.036)	2.3	2.6	4.0	4.2
η	(1/2)	2.7	2.8	4.0	4.4
r_0	(0.09)	2.4	2.7	6.8	2.8

a. The numbers in this column are the baseline parameter values.
b. The variation in α is only 25 percent.
c. In these cases the reported measurement error is larger than the maximal real gap as reported in table 9.2.

9.6 Summary and Conclusion

This chapter has examined some of the reasons why growth in the measured level of real output and real income might fall with the introduction of a new general purpose technology that raises the productivity of R&D and hence draws resources out of producing goods into producing knowledge. It has focused on two particular mechanisms, one that could cause an illusory slowdown of GDP and one that could cause a real slowdown. The reason why an illusory slowdown might occur is that national income accounts to a better job of measuring the output of goods than of knowledge; this is what I call the knowledge-investment problem. The reason why a real slowdown might occur is that a faster pace of R&D will raise the rate of capital-obsolescence, through Schumpeter's process of creative destruction. This will in turn cause a decline in the capital stock per efficiency unit of labor large enough to offset the increased rate of technological progress coming from the rise in innovative activity.

The chapter constructed an R&D-based endogenous growth model and calibrated it to the U.S. economy in order to give an idea of the orders of magnitude involved with these effects. It found that while both factors were empirically significant, the real factor of obsolescence was quantitatively more important than the illusory factor of measurement error. Even if there were no measurement error at all (if R&D output were completely capitalized), the growth rate would be reduced for 14 years following the introduction of the GPT, and the level of real output per person would be lower than it would have been without the introduction for 28 years.

The knowledge-investment problem in measured GDP will exaggerate both the size and duration of the slump. Measured GDP falls below its no-shock reference path by an amount that reaches 8.5 percent after 22 years, whereas the maximum shortfall in actual GDP is only 6.6 percent and occurs after 15 years. But, as these numbers indicate, the illusory component of the slump is small relative to the actual slump caused by accelerated obsolescence.

In fact the size of the slump in *national income* is probably *under*stated by the national accounts because of the difficulty of adjusting the depreciation rates used in the accounts to reflect changing rates of obsolescence. As the chapter's numerical model shows, the effect of introducing the new GPT is a rise in the rate of obsolescence by so much that national income, properly measured, falls by 8.1 percent in the first year. If, however, no adjustment is made to depreciation rates in calculating national income that year, as seems likely, the measured level of national income falls by only 1.9 percent (because of the illusory effects of the knowledge-investment problem).

An important implication of this analysis is that the relationship between R&D and economic growth may involve a much longer lag than time-series econometricians are used to dealing with. The rise in R&D that takes place at the onset of the above simulation and lasts for 23 years does not yield an increase in the growth rate until it has been underway for 15 years. Even then, as is clear from figure 9.7, the increase in growth does not become pronounced until R&D drops, in year 23. Thus a casual observer could easily be misled into thinking that R&D hampers growth. When the lag between cause and effect is this long, causal relationships are particularly difficult to discern, and even carefully formed judgements are prone to error. The best defense against such errors is a good theoretical model of the interrelationships between R&D and economic growth.

Appendix A

This appendix verifies that both steady-state loci in figure 9.1 shift down when λ increases and that the downward shift in the $\dot{k} = 0$ locus is larger than in the $\dot{v} = 0$ locus, as indicated in figure 9.2.

From (18), the equations of the two curves are

$$sk^\alpha - [\delta + n + \phi(v, \lambda)(1 - \eta + \sigma)]k = 0 \quad (\dot{k} = 0),$$

$$-\alpha(1 - \alpha)q^{-1}k^\alpha + [\alpha^2 k^{\alpha-1} - \delta + \eta\phi(v, \lambda)]v = 0 \quad (\dot{v} = 0).$$

Hence

$$\left.\frac{\partial v}{\partial \lambda}\right|_{\dot{k}=0} = -\frac{\phi_\lambda}{\phi_v} < 0$$

and

$$\left.\frac{\partial v}{\partial \lambda}\right|_{\dot{v}=0} = -\frac{\eta\phi_\lambda v}{\eta\phi_v v + \alpha(1 - \alpha)q^{-1}k^\alpha/v} \in \left(-\frac{\phi_\lambda}{\phi_v}, 0\right).$$

That is, both curves shift down but the $\dot{k} = 0$ curve shifts down by more.

Appendix B

This appendix shows how the implied values in table 9.1 are derived.

Recall that $\beta_0 = (1 - \eta)\phi_0$ and $g_0 = \sigma\phi_0$. From this,

$$\phi_0 = \frac{\beta_0}{1 - \eta} \quad (\phi_0),$$

and

$$\sigma = \frac{g_0}{\phi_0} \quad (\sigma).$$

These two equations, together with equations (14) and (16) and the assumption that we are starting in a steady state, imply that

$$sk_0^{\alpha-1} = \delta + n + \beta_0 + g_0$$

and

$$\alpha^2 k_0^{\alpha-1} = r_0 + \delta + \beta_0,$$

from which the saving rate is

$$s = \alpha^2 \frac{\delta + n + \beta_0 + g_0}{r_0 + \delta + \beta_0} \quad (s),$$

and the initial capital stock per efficiency unit is

$$k_0 = \left(\frac{r_0 + \delta + \beta_0}{\alpha^2}\right)^{1/(\alpha-1)} \quad (k_0).$$

As explained in the text, the number of products per person is

$$q = \frac{\xi}{n} \quad (q).$$

According to the production function (7), initial GDP per efficiency unit is

$$y_0 = k_0^\alpha \quad (y_0).$$

From (12) and the assumption that we are starting in a steady state, the initial productivity-adjusted value of an innovation is

$$v_0 = \frac{\alpha(1-\alpha)q^{-1}k_0^\alpha}{r_0 + \phi_0} \quad (v_0).$$

By definition, the initial level of R&D expenditure is $p_0 Q_0 A_0^{\max}$. According to (10) this can be rewritten as $\gamma \phi_0 v_0 Q_0 A_0^{\max}$. Hence initial R&D as a fraction of GDP can be expressed as

$$\left(\frac{R\&D}{Y}\right)_0 = \frac{\gamma \phi_0 v_0 Q_0 A_0^{\max}}{y_0 N_0 A_0} = \frac{\gamma \phi_0 v_0 q (1+\sigma)}{y_0}.$$

This can be solved for the R&D elasticity of innovations:

$$\gamma = \frac{y_0 (R\&D/Y)_0}{\phi_0 v_0 q (1+\sigma)} \quad (\gamma).$$

Finally, from equation (17), the initial value of the productivity parameter λ in the research technology is

$$\lambda = \phi_0^{1-\gamma} (\gamma v_0)^{-\gamma} \quad (\lambda).$$

References

Aghion, P., and P. Howitt. 1992. A model of growth through creative destruction. *Econometrica* 60: 323–51.

Aghion, P., and P. Howitt. 1996. The observational implications of Schumpeterian growth theory. *Empirical Economics* 21: 13–25.

Arroyo, C. R., E. Dinopoulos, and S. G. Donald. 1994. Schumpeterian growth and capital accumulation: Theory and evidence. Unpublished. April 1994.

Baily, M. Neil, and R. J. Gordon. 1988. The productivity slowdown, measurement issues, and the explosion of computer power. *Brooking Papers on Economic Activity* 2: 347–420.

Benassy, J.-P. 1998. Is there always too little research in endogenous growth with expanding product variety? *European Economic Review* 42: 61–70.

Caballero, R. J., and A. B. Jaffe. 1993. How high are the giants' shoulders: An empirical assessment of knowledge spillovers and creative destruction in a model of economic growth. In *NBER Macroeconomics Annual*. Cambridge: MIT Press, pp. 15–74.

David, P. 1990. The dynamo and the computer: An historical perspective on the modern productivity paradox. *American Economic Review* 80: 355–61.

Freeman, C., and C. Perez. 1988. Structural crises of adjustment, business cycles and investment behaviour. In G. Dosi, C. Freeman, R. Nelson, G. Silverberg, and L. Soete, eds., *Technical Change and Economic Theory*. London: Pinter, pp. 38–66.

Gordon, R. J. 1990. *The Measurement of Durable Goods Prices*. Chicago: University of Chicago Press.

Griliches, Z. 1994. Productivity, R&D, and the data constraint. *American Economic Review* 84: 1–23.

Helpman, E., and M. Trajtenberg. 1994. A time to sow and a time to reap: Growth based on general purpose technologies. CIAR Working Paper 32.

Howitt, P. 1996. On some problems in measuring knowledge-based growth. In P. Howitt, ed., *Implications of Knowledge-Based Growth for Micro-Economic Policies*. Calgary: University of Calgary Press.

Jones, C. I. 1995. R&D-based models of economic growth. *Journal of Political Economy* 103: 759–84.

Katz, L. F., and K. M. Murphy. 1992. Changes in relative wages, 1963–1987: Supply and demand factors. *Quarterly Journal of Economics* 107: 35–78.

Kortum, S. 1993. Equilibrium R&D and the patent-R&D ratio: U.S. evidence. *American Economic Review* 83: 450–57.

Lipsey, R. G., and C. Bekar. 1995. A structuralist view of technical change and economic growth. In T. J. Courchene, ed., *Technology, Information and Public Policy*. 9–75. The Bell Canada Papers on Economic and Public Policy, vol 3. Kingston: John Deutsch Institute for the Study of Economic Policy, pp. 9–75.

Mankiw, N. G. 1995. The growth of nations. *Brooking Papers on Economic Activity* 25: 275–310.l

Young, A. 1995. Growth without scale effects. Boston University Working Paper.

10   The Division of Inventive Labor and the Extent of the Market

Timothy Bresnahan and
Alfonso Gambardella

10.1 Introduction

Vertical disintegration of economic activity is the core of the division of labor as Smith (1776) and Stigler (1951) saw it. Stigler's conjecture was that *general specialties* were the economically important examples of division of labor. Not an oxymoron, the phrase has instead Stiglerian compactness and precision. A division of previously unified labor can create a "specialty." That specialty will be of economywide importance when its scope of application is "general," meaning that it is used by many kinds of customers. In the first and second industrial revolution, many general specialties exploited scale economies in production. Stigler thought of transportation examples like railroads and shipping, of financial examples like the London banking center, and of specialized production of intermediate materials (steel, chemicals) or capital goods (machine tools, electric motors, and lights). In modern times the creation of new general specialties continues unabated. Specialized science and engineering-based high-tech industries, broadly useful through the economy, lead this trend. Electronics, for instance, has been accompanied by sustained increase in specialization as hardware, software, networking, and so on, have become separate engineering subdisciplines practiced by specialists working in separate industries. Yet these are general specialties, selling the fruits of their invention to a wide variety of distinct types of customers. In short, both today and in the past, one of the most apparent effects of the division of labor has been the creation of whole new bodies of specialized

---

We thank the Alfred P. Sloan Foundation and Italian Consiglio Nazionale delle Ricerche under contract number 94.02070.ct10 for support, an editor of this volume, a referee, Ashish Arora, Victoria Danilchouk, Nathan Rosenberg, and Manuel Trajtenberg for comments and discussion.

knowledge and frequently whole new industries selling to many others in the economy.[1]

These general specialties, or GPTs, are an important source of economy-wide scale economies and growth (Bresnahan and Trajtenberg 1995; Helpman and Trajtenberg 1995).[2] Moreover the fact that GPTs, or more generally scale economies in invention, can drive growth, is a clear implication of the new growth theory (Romer 1986, 1990). But while the causal link from division of labor to growth through GPT seems clear, the modern theoretical literature has cast some shadows on how growth and the attendant increase in scale economies permit exploitation of the division of labor. Smith and Stigler argued that division of labor is limited by the extent of the market. By contrast, the main theorem in the modern literature shows that a large market will lead to vertical integration, *not* vertical disintegration and specialization (Perry 1989, sec. 7.3.) The literature has concluded that the problem lies in the Smith-Stigler treatment of *specialization*. In order to obtain a theory in which the Smith-Stigler conjecture is correct, it has then proposed a definition of specialization based not on scale economies but on scope *dis*economies.[3] While this saves the theory formally, it has the very unhappy implication that division of labor is not a mechanism for positive feedback from growth through scale economies to more growth but rather a mechanism for ameliorating some of the costs of growth under *convex* (no scale economies) technology.

Another literature, in growth theory, has changed the definition of "specialization" in a different way in order to get a result with a Smith-

---

1. Despite the fact that his famous example of the "pin factory" concerned the division of labor across workers within the firm, Smith himself clearly envisaged a much broader division across industries or activities based on deep and generalized knowledge bases. He first argued that improvements in machinery are sometimes made by philsophers or men of speculation: "who ... are often capable of combining together the powers of the most distant and dissimilar objects." Then he noted that "... in the progress of society, philosophy or speculation becomes, like every other employment, the principal or sole trade and occupation of a particular class of citizens ... it is subdivided into a great number of branches ... and this subdivision of employment in philosophy, as well as in every other business, improves dexterity, and saves time." (Smith 1776, ch. 1)

2. We will treat "general specialty" and "general purpose technology" as nearly identical in meaning. To the extent that there is a distinction, a GPT's body of knowledge or the people who know it are a general specialty.

3. "The difficulty with the Stigler model occurs from trying to capture the notion of specialization.... An alternative view of specialization is that there are economies from doing a limited set of activities, rather than economies of scale from simply doing a lot of any one activity.... Specialization would mean diseconomies of scope across vertically related production processes" (Perry, 1989, p. 232). Perry directs our attention to the agency costs which might cause the scope diseconomies, a direction that the information-based theory of the firm has followed up carefully (Milgrom and Roberts 1992.)

Stigler flavor. Using monopolistic competition models, papers such as Romer (1992) show that a larger economy will have more product variety as the costs of specialized inputs can be spread out over more units. In this way they show that large market size causes *horizontal* specialization, a division of labor among firms or workers that make substitute goods. These models are obviously right, and the increasing product variety of modern consumer economies is clearly an important part of the growth process. Yet this analysis is unrelated to the *vertical* "specialization" and "division of labor" noted by Smith and Stigler and so important in modern high-tech industry.

In this chapter we argue that the problem does not arise from an incorrect definition of specialization but lies instead with the modern theory's definition of the *extent of the market*. Vertical integration theories have formalized the extent of the market as the output of a single good, asking when two production tasks will be vertically disintegrated. Our alternative definition views vertical disintegration as the introduction of a general specialty and the founding of a GPT. The point is that the extent of the market for a GPT is not only the volume of production in one of the sectors that applies it but also the number of distinct applications sectors that might apply it. Taking the extent of the market to mean breadth as well as depth connects the theory to the world. The general specialties noted by Stigler all had a distinctly "infrastructural" flavor, as do many of the modern GPTs. Their customers are widely dispersed through the economy, using their products, services, and technologies in a wide variety of distinct ways.

In our theory the forces leading to the division of inventive labor are quite simply those that make a GPT part of the cost-minimizing organization of inventive effort. These forces call for general specialists, and the inventions by these specialists are facilitated by economies of scale because, within the specialty, there is sharing or re-use of concepts and tools. But the other important point is that the broad applicability of a general specialty is not free. An inherent tension in any division of labor is that the distinct users of a technology, or for that matter of a good or service, employ it for different purposes. Consequently they have different needs, and these needs would be best satisfied by producing, adapting, or using the technology or input according to their special goals and demands. This is a force for localization. Standardizing the technology or input allows exploitation of the gains from specialization, while localizing it permits superior matching. A general specialty is a compromise between the scale economies inherent in specialization and the failure to localize inherent in generality.

In the next section we build a very simple model of this trade-off. The only open issue is whether to create a general specialty input or a set of localized specialty inputs. The model has increasing returns in the specialty, a cost of mismatching, and no other elements. In section 10.3 we show how two different definitions of "the extent of the market" lead to opposite results for vertical specialization. Increasing the size of all applications sectors leads to vertical integration, while increasing the variety in applications sectors leads to division of labor. This model is the simplest one in which the Smith-Stigler conjecture is correct and can be distinguished from the confusing modern alternative. The existence of a general purpose technology in this model is endogenous in equilibrium.

In section 10.4 we examine several contemporary and historical GPTs. This leads us to, in section 10.5, a second model with extra features that we think capture some of the essence of science- and engineering-based general specialties. In particular, the second model has an extremely simple treatment of how science's abstract and explicit representation of knowledge becomes useful. The second model suggests a higher economic return to formal science and engineering in a larger and more diverse economy because generality in the representation of knowledge itself lowers the costs of broad and general use. The main result of the model, reported in section 10.6, is that the same force, broad application, that leads to general specialties is also a force for abstraction, science, and formal engineering. Just as the division of labor is limited by the extent of the market, so too is the division of inventive labor. In the last section we examine some very specific cases of general specialty industries. Extensive markets in the "many-uses" sense are a far better predictor of the division of inventive labor than are extensive markets in the "large-uses" sense.

We argue that our first model explains the role of general specialists, at an abstract level, in economic growth generally. Like any scale economies model, it faces the very real challenge of explaining why the scale economies are not exhausted in large economies. Our second model shows how opportunities for increasing the formality, abstraction, and generality of scientific and engineering knowledge have preserved an economic role for GS scale economies. Exploitation of these opportunities has called for further specialization of knowledge creators, further vertical disintegration of inventive effort, and the creation of new GPTs. Two important forces are at work here. First, science and engineering have been amenable to specialization, refinement, and formalization. Smaller and smaller subspecialties are created, which are potentially but by no means necessarily available for commercial application. Second, modern economies are

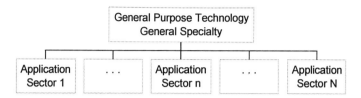

Figure 10.1
Industry structure with a GPT

diverse and complex as well as large. Consumerist societies have a wide variety of kinds of products and services, and international trade (supported by better transportation and communication) means that a number of distinct societies can be served by the same technology. These two forces imply that new general purpose technology industries can be founded and can evolve into new engines of growth.

10.2 General Specialists and Cost-Minimizing Industry Structure—The Simplest Model

We consider an activity, like the creation of new knowledge or the provision of intermediate inputs, which has a tension between generality and localization. The knowledge or inputs are broadly useful and scale economies in their creation or provision are important, so there is a value to generality. Yet different uses need different varieties of the knowledge or features of the input, a force for localization. In this context, we consider two alternative systems of organization of economic activity, shown in figures 10.1 and 10.2.

The first has a general purpose technology based upon a general specialty. The industry structure is vertically disintegrated, with a GS-based GPT selling to many application sectors (AS) sectors each of which performs its own localization. In the other, there is complete localization of invention within each AS. No independent GPT sector exists. Existing theory has done a good job of understanding the incentive, welfare, and growth implications of a GPT but has not yet explained when a GPT will come into existence. Equilibrium in our model will determine which of the two systems, one with, the other without, a GPT, will organize activity.

Abstracting away from incentives issues, we ask which system minimizes costs. This leads immediately to the (true) Smith-Stigler theorem with the right definition of "the extent of the market" and also to the counterintuitive modern theorem with the other definition.

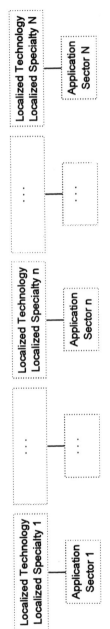

Figure 10.2
Industry structure with localization

10.2.1 Setup of Problem and Assumptions about Costs

Many of the elements of our very simple model are the same under either system. We posit N distinct uses of the knowledge or input—applications sectors. Each use has a desired rate of production, Q_n. Since ours is a theory of cost, we take N and Q_n to be given and exogenous.

Each sector needs to undertake two activities, which might be understood as production processes or as inventions. One of the two activities is always localized and serves to define the identity of an AS. The other activity might be localized or might be served by a general specialty. These have cost functions given by $c_n(Q_n)$ and $K(Q)$, respectively. We make no assumptions about the $c_n(\ )$, since they do not affect our results. The costs $K(\ )$ are the costs of providing the potentially general input. We assume that there are increasing returns in this activity—that is, the elasticity of $K(\ )$ is smaller than 1. Moreover we assume that the degree of scale economies is falling with the rate of output, namely $K(Q)/Q - K'(Q)$ declines with Q. These assumptions on the shape of $K(\ )$ are crucial.

10.2.2 Costs under Localization and General Specialty

If there is no general specialty, the costs in sector n are given by

$$c_n^L = c_n(Q_n) + K(Q_n), \tag{1}$$

where the superscript L refers to localization. The first term is the costs of the activity that occurs only in this AS. In this localized industry structure the costs of providing the potentially general input are borne separately by each use.

If there is a general specialty (a GPT sector), costs in it are

$$C^{GS} = K\left(\sum_{n=1}^{N} Q_n\right). \tag{2}$$

Note that the function $K(\ )$ of the GPT sector is the same as for each individual AS under localization. However, they will be lower than the corresponding costs in (1) because of the scale economies inherent in $K(\ )$.

Under a general specialty system, however, costs in the applications sectors are raised by "mismatches" and absence of variety. We represent these costs as

$$C_n^{GS} = c_n(Q_n) + d_n Q_n. \tag{3}$$

Here $d_n \geq 0$ is the "distance" of the applications sector from the general input or knowledge. The greater this distance, the greater the penalty for failure to localize the input or knowledge. Our model is one of overall cost minimization, so we ignore transfer prices, contracts, and so on.

10.3 Division of Labor and Extent of Market, I

We now examine the conditions under which our first model will have division of labor in cost-minimizing industry equilibrium. We examine the comparative statics of cost-minimizing industry structure as we increase the extent of the market. We use two definitions of the extent of the market: Q_n the desired production in each use, and N the number of uses. For simplicity we offer proofs only for the symmetric case where all N sectors have the same $Q_n = Q$ and the same $d_n = d$.

The cost-minimizing structure is general specialist whenever

$$\sum_n (c_n(Q) + dQ) + K\left(\sum_n Q\right) < \sum_n (c_n(Q) + K(Q)).$$

In the vertical integration literature the theorem below contradicts the Smith-Stigler conjecture.

Theorem 10.1 Increases in the size of each use, Q, can change cost minimizing industry structure from general specialist to localized and vertically integrated but not the other way.

Proof We differentiate the difference in costs between the two structures with respect to Q and evaluate the derivative at a point where the two regions are equal cost (if such a point exists). Define Δ to be the difference between industry costs under GS and localization. Then $\Delta \equiv NdQ + K(NQ) - NK(Q)$, and

$$\left(\frac{\partial \Delta}{\partial Q}\right) = Nd + NK'(NQ) - NK'(Q).$$

At $\Delta = 0$, $Nd = N[K(Q)/Q - K(NQ)/NQ]$. Replace this in the expression for $(\partial \Delta/\partial Q)$. One obtains

$$\left(\frac{\partial \Delta}{\partial Q}\right) = N\left[\frac{K(Q)}{Q} - K'(Q)\right] - N\left[\frac{K(NQ)}{NQ} - K'(NQ)\right].$$

But the degree of economies of scale is falling with the rate of output. Then $K(x)/x - K'(x)$ declines with x, and therefore $(\partial \Delta / \partial Q) > 0$. ∎

This theorem goes the reverse way of the Smith-Stigler theorem because it has a local definition of the extent of the market. As each use grows large, the fixed costs of inventing technology for it grow small by contrast. But the costs of using inappropriate technology or infrastructure scale up along with the market. This is by far the most common result in the literature. It is almost always demonstrated for the $N = 1$ case because large Q is the only definition of "extent of the market" considered in the papers.[4] It is obvious that the result extends to such alternative comparative statics as increasing the Q_n of a single sector—that sector can switch from but not to using the GPT. Similarly extensions of the theorem to the case of heterogeneous d_n and $K_n(\ )$ call only for a treatment of the possibility that some AS's will use the GPT, while others will not but will have the same kinds of results as here. We now show a result in which the Smith-Stigler conjecture is correct.

Theorem 10.2 An increase in the number of uses, N, can change cost-minimizing industry structure from localized and vertically integrated to general specialist but not the reverse.

Proof By same method evaluate $(\partial \Delta / \partial N)$ at $\Delta = 0$. One obtains

$$\left(\frac{\partial \Delta}{\partial N}\right) = -Q\left[\frac{K(NQ)}{NQ} - K'(NQ)\right] < 0. \quad \blacksquare$$

This result is different because the margin of more uses (N) is economically quite different from the margin of larger uses (Q). Larger uses tend to call forth more localized and locally optimized inputs. More, or more diverse, uses call for the creation of general specialties, as the general specialist can be spread out over more units. This highly stylized model cannot reveal very much more than the distinction between the two margins, but it does show that distinction very clearly.

These two results clarify an important aspect of the Smith-Stigler conjecture that division of labor is limited by the extent of the market. If by the latter one means the number of firms of a given size, then division of labor is encouraged by an increase in the number of such firms. But, if increases in the size of the market occur through increases in the size of firms or sectors (i.e., in the potential demand of applications of a certain type), there is a discouraging effect on division of labor. The intuition is

---

4. See Perry's (1989) discussion.

straightforward. Suppose that the market expands because of a larger potential demand for the particular good produced by each firm. These firms can then spread the fixed cost of producing a localized technology on a larger internal market, which encourages them to invest in the local technology. But, if the market expands because of increases in the number of potential uses of the GPT, then the value of a *general* solution rises.[5] Thus only under the latter definition of the extent of the market is the Stigler-Smith conjecture true. This is clearly what Stigler had in mind. The infrastructural industries he saw as general specialties provided a well-defined service or product to all comers in a market organization.

10.4 Division of Labor and Resulting Economies in Several Industries

We now examine the real world of several historical and contemporary general specialty industries. This scrutiny shows two things. First, our assumptions about $K(\ )$ and d_n have different but related analogies in the real world in each industry, so that the logic of the creation of a general specialty emerges clearly. Second, once each general specialty was created, technologists and business people noted the rigidities associated with the exploitation of scale economies. This led to attempts to lower d, namely to make the general specialty more adaptable and widely useful. The importance of this kind of change, and the way it drew on formal science and engineering, are an important part of our second model.

For Stigler's example of the railroad, long-run increasing returns to scale have two distinct meanings. First, the physical capital of a railroad is efficiently shared across many classes of shipments. Second, the (very considerable) *invention* costs of improvements in steam power, steel rails, telegraph, and management structures in order to control large transport systems could be similarly spread out. The cost for any user of building dedicated transportation lines linking, say, a particular shipper's most habitual routes could be considerable. Relatedly, single shippers were unable to generate enough transportation demand to justify these setup costs.

The corresponding d's are not zero. Shippers of different kinds care differently about speed, reliability, smoothness, and cost. A railroad opti-

---

5. Earlier work which has anticipated our point has noted the value of the generality—see, for example, Rosenberg (1963) on technological convergence—or seen the general market as a source of even more returns to scale than the largest specific market—see, for example, Romer (1996)—without noting that the emergence of a GPT itself is contingent on the economic return to broad rather than deep application.

mized to deliver fresh fruit or passengers differs from a coal or grain carrier. By relying on specialized suppliers, the users were giving up the opportunity to ship their freight at their most desired moments in time, along the optimal routes between departure and destination, or under any other very special condition. This happened for the obvious reason that the layout of the railroad system, and its scheduled routes and timing, had to be optimized according to the utilization of the network as a whole by its many customers rather than fitting the exact needs of individual users. Nonetheless, the combination of operational and invention scale economies swamped these modest benefits of diversity and a general specialty, railroading, emerged.

10.4.1 Making the General Specialty More Adaptable

In the eighteenth and nineteenth centuries, when railroads were a new and high-technology industry, a growing demand for railroad transportation by many customers with different shipping requirements (timing, destination, routes, etc.) gave a great impetus to improve and widespread coordination within and among railroad companies, thereby improving the efficiency of utilizing the network by any of its users. As Chandler (1990, pp. 53–56) points out, this was carried out by notable organizational and technological innovations. The U.S. railroad companies pioneered the techniques of modern management. They undertook considerable steps toward a "scientific" approach to the scheduling of movements of trains, freight, and passengers; and to the optimization of routes and connections between hundreds of locations and destinations; and to the maintenance of railroads and related equipment. This very exact scheduling, which was critical to enhancing the efficiency of transportation, was created by subdividing vast and very complex set of operations into a hierarchy of smaller and simpler tasks, which were supervised, monitored, and coordinated by different layers of managers. The effect of this advance was to progressively lower, but never to remove, the d costs associated with railroads as a transport system.

Fundamental change in the conditions of localization for transport awaited the invention of the automobile and the truck.[6] This technology shifted the boundary between the general and the localized. "Road" continued to be general (and indeed shifted to public provision), whereas "rolling stock" and "management" became specific to the using sectors.

---

6. See Bresnahan and Gordon (1996) for a discussion.

Now a user would own its vehicles and, subject to a congestion externality, schedule its own shipments or travel. Even though motor vehicles are subject to vastly less scale economies than are trains, they have increased flexibility, breaking the rigidity of sameness and standardization. A less steeply sloped $K(\ )$ permitted considerable escape from generalist production and from the d costs imposed on users.

A different example comes from the twentieth-century chemicals industry. At the beginning of the century, each firm designed its own manufacturing process. There was very little sharing of process knowledge across makers of different products. The emergence of the chemical engineering discipline changed that radically. As noted by Rosenberg's (1997) chapter in this book, chemical engineering emphasizes the quantitative analysis of basic "unit operations" such as distillation, evaporation, drying, filtration, absorption, and extraction, and this provided a unified framework for thinking about the design of different processes in oil refining and in chemical production. While analysis of these operations varies with the material being operated upon, the analytical principles do not vary. The engineering discipline, with strong roots in science, was able to advance understanding of the general analytical principles and of their mode of application to different materials. Thus the invention of a general specialty involved a division between the general (here process knowledge) and the specific (the application of that process knowledge to particular products).

Most important, this development created enormous opportunities for specialization of the invention function itself. Especially with the high growth of chemical and oil refining markets immediately after World War II, the industry witnessed the formation of many companies with deep process knowledge, the specialized engineering firms (SEFs).[7] Their services assisted chemical and petrochemical companies in the design and engineering of chemical plants and oil refineries. The SEFs supplied process design and engineering services for a wide number of products (plastics, fibers, elastomers, etc.). Some of the most prominent SEFs have generated important *process* innovations, and significant improvements of existing processes.[8] Universal Oil Products (UOP), for instance, has been responsible for a tremendous number of inventions throughout this century. UOP acted as the R&D department of many small and independent oil refiners and chemical firms. Still today it holds a huge number of

---

7. This story is discussed in detail in Arora and Gambardella (1997).
8. See Arora and Gambardella (1997). See also Freeman (1982).

licenses in many oil refining and chemical processing technologies, and in several countries.

Two of UOP's technologies figure as quintessential examples of general purpose inventions—the first continuous cracking process for producing gasoline, the Dubbs process, developed in the 1910s, and the Udex process for separating aromatic chemical compounds from mixed hydrocarbons, developed in the 1950s. The value of the Dubbs process was twofold. It worked continuously, without stopping production. And it produced gasoline from either high-quality feedstocks or from low-quality "black oil." "With the Dubbs process, UOP could live up to the 'universal' in its name by cleanly cracking *any* oil, regardless of coke-formation quantities."[9] Rather than vertically integrate forward into refining, UOP licensed its technology to the myriad local refiners, helping them specialize it to their particular feedstocks.[10]

During the 1950s UOP developed the Udex process to separate aromatic compounds (benzene, toluene, and xylene) from mixed hydrocarbon streams. These aromatic compounds are themselves general purpose inputs used in the making of many distinct chemical prospects. The Udex process was extremely flexible: "Generally, UOP has been able to assemble a combination of processing 'blocks' that would allow a producer to make any desired combination and relative quantity of benzene, toluene, and xylene isomers from every conceivable feedstock."[11]

These examples show that the SEFs' expertise was specialized in the sense that it was deep in particular processes but general in the sense that it cut across many products. Further the SEFs were able to generalize their process knowledge to a wide variety of distinct feedstocks.

Parallel to the SEFs, university chemical engineering departments played a crucial role in making fundamental process improvements. Many fundamental process improvements were made in universities or SEFs and shared across a wide variety of firms making a wide variety of products. Here the source of increasing returns in $K(\ )$ lies entirely in the *invention* of improved processes. Once the chemistry was understood, it was wasteful to separately advance knowledge of how to improve the same process for distinct chemical products. Further the general inventions lowered the costs of particularization, namely lowered d. The general

---

9. Remsberg and Higdon (1994, pp. 50–51). Italics in the original.
10. Note the incentives that arose because of a large market in the "many-uses" sense: "UOP's approach to process licensing was particularly applicable to the oil refinery business of that era, Practically every little town in the country with access to oil had a small refinery ...." (Remsberg and Higdon 1994, p. 50).
11. Spitz (1988, p. 191).

specialists in universities and in the SEFs made inventions of general scope and value, with the knowledge represented in the manner of general scientific principles. Applications to local circumstances were just applications, analogous to students' problem sets rather than to the development of a whole new body of knowledge.

One feature of the chemicals history is that things that had been done distinctly came to be done in the same general way. Rosenberg (1963) identifies a process of technological convergence that creates general specialties. While the distinct applications sectors had been separately served by distinct technologies (e.g., industry specific machine tools), after convergence they are served by a common, general input. Our theory ignores this dynamic, only suggesting that large and diverse markets increase the return to this kind of generalization.

A related example is the invention of the microprocessor by Intel in 1971. Before then, integrated circuits were largely "dedicated" products, in the sense that their operations were defined by the physical wiring and interconnections designed and built by the manufacturer on the chip. An important consequence was that the circuits had to be produced by the manufacturers with specific applications in mind. By contrast, the microprocessor, or "programmable chip" as it was aptly named, could be programmed. As Braun and Macdonald (1982) note, it implied "software wiring" as opposed to "hardware wiring." It could then read and process more variable instructions, and perform a far larger number of operations. Most important, it could be produced without specific applications in mind. Its functions could be defined to a greater extent by the users themselves who could program the chip according to their needs.[12]

It is then not surprising that the device rapidly found extensive applications. Apart from its core use in microcomputers, it became a pivotal component in telecommunications, aerospace, and office equipment, in the control of industrial processes, in the automobile industry, among many others. Its utilization extended the range of applications of integrated circuits. In a sense this widespread application was the natural consequence of the fact that whether deliberately or not, the microprocessor was conceived, from its very invention, as a general purpose object. The impact

---

12. Braun and Macdonald (1982) suggest that the distinction between the microprocessor and the earlier integrated circuits is really that between a computer and a calculator. The latter can perform only the functions that are "permanently" defined on its chip and that can be activated by pressing special keys (e.g., numbers or arithmetic operations). The computer instead can read instructions defined in many possible ways (logic, arithmetic, etc.) and therefore perform more elaborate and distinct operations. They also note some intermediate forms, like the erasable programmable read-only memory.

was to lower d. Ultimately the microprocessor also changed part of the integrated circuit business into one characterized by very steep $K(\ )$. A general microprocessor is a very complex device, and we now see a small number of firms making very long production runs of a few microprocessor designs.

The uses of integrated circuits vary along another dimension as well, the performance–cost trade-off. Hardware wiring, despite all the wasteful duplications of design costs (multiple $K(\ )$) it involves, offers superior performance in many applications. An "applications-specific integrated circuit," or ASIC industry, flourished in parallel to the microprocessor GS. Here the division of inventive labor is quite different. Manufacture of ASICs is performed by general specialists. But unlike Intel or Motorola, these are specialists in the manufacturing process only. They do not design the products they make. Applications sectors design ASICs and solicit manufacturing cost bids from these general specialists. A fundamental organizational innovation has arisen to lower the d costs in this industry. A language has emerged for describing ASIC designs. It is a computer language, spoken by two very different kinds of computers. The first are computer-aided design workstations used in the AS. The second are manufacturing-control computers used in the GS. By this mechanism even ASICs can have "software wiring."[13] The AS firm designs a logical chip, and the GS firm makes it. Thus both the substantial scale economies in the plant (steep $K(\ )$) and the substantial benefits of localization have been achieved by this d-lowering organizational invention.

In sum, the more important modern general specialties have been in the science- and engineering-based industries. These offer a distinct advantage, one in which the tension between localization and generalization is lessened by invention of lower d ways to organize inventive activity. The advantage arises in the use of regularized, systematized knowledge to make AS invention easier. This works to draw more AS's into the ambit of the GPT. Our model needs an extension to deal with these greater opportunities, but an extension that is consistent with the basic thrust of the model.

10.5 Model with Endogenous Generality and AS Voluntarism

We add two features to our model in order to capture the tension between localization and scale economies more fully. First, we model the

---

13. That is, they can have software wiring at design time if not at use time. ASICs are more suitable for use in a wide variety of special-purpose devices than in, say, computers.

general specialty as being able, for a fixed cost, to lower d. This captures the costs and benefits of a more abstract, portable, and reconfigurable technology, as described in the examples in the last section. We will interpret lower d equilibria in our model as the endogenous creation of a more general purpose technology. Second, we permit the AS to be of different sizes and to substitute in and out the use of the GPT. Then there can be variety in the technology choice of the AS's. Some users (the larger ones, from the analysis of the earlier section) develop a localized version of the generally useful technology. Others, and the boundary here will be endogenous to equilibrium, can choose to participate in the general purpose technology, using the goods or services of the general specialist. This involves two changes: We differentiate between the size of the market in the AS and the size of the market the AS offers to the GPT, and we permit sectors to vary in their participation.

10.5.1 Setup of Model: Localized or General Purpose Technology

The AS sectors produce a final good using a potentially general intermediate input whose quantity is still denoted by Q. We now add the size of the market demand for the specific good produced by each sector, denoted by S, and we assume that S is distributed across firms according to a density function $f(S)$. Thus our sectors are exogenously of different sizes, though the size of their demand for the general input is endogenous.[14]

The gross surplus in the AS is $\pi(S, Q)$. Apart from the obvious assumptions that $\pi_S > 0$ and $\pi_Q > 0$, we assume that the cross-partial of π with respect to S and Q is positive, $\pi_{SQ} > 0$. A larger sector obtains larger gross benefits from Q. There are underlying price and quantity decisions about the final good in the background, but they can most safely remain there.[15] We focus attention on the demand for Q and the technology choice of the AS as a function of S and of the GS' offerings.

We are interested in two experiments. The first experiment consists of increasing the *number* of applications sectors, which is accomplished by replication, increasing the density $f(S)$ proportionately at all S. The second

---

14. We will refer to S as being the size of the market of each sector or more simply the size of the sector. To save notation, we will not use the subscript n for firms when referring to S and Q.

15. Equilibrium in the AS maximizes $\pi(S, Q)$. Because of our theory's strong cooperative flavor, it is natural to interpret it $\pi(S, Q)$ either as the profits of the AS, if the AS is supplied by a monopoly firm, or as the sum of consumer plus producer surplus. See Bresnahan and Trajtenberg (1995) for more interpretation.

experiment consists of increasing the *size* of all applications sectors. This is accomplished by scaling the size of the sectors to be $S^* = \alpha S$, instead of just S. We assume that S is still distributed according to the density function $f(S)$, and that α is a proportionality factor.

Localized or GPT Choice for a Single AS

Once again, the input Q can be of two types. On the one hand, the AS can choose to utilize a "local" technology. That is, they develop a specialized technology at their own location, and this can be designed to suit in a fairly exact way their special needs. If the sector uses a local technology, they incur costs $K(Q)$ to generate it. The net surplus is

$$\Pi^L \equiv \pi(S, Q) - K(Q),$$

where the superscript L denotes the "localized" technology regime. We also define $Q^L(S)$ to be the optimized Q under this regime and $\Pi^L(S)$ to be the corresponding optimal profits. Our assumption on the cross-partial of π implies that both $Q^L(S)$ and $\Pi^L(S)$ increase with S.

On the other hand, the sector can choose to buy the input Q from the GPT sector (if one exists). In this case they do not incur $K(\ )$, but they incur the d costs and a cost w per unit of Q, which is the price of the GPT. The total unit price of using the GPT will be $(w + d)$. The payoff to the AS in the "generalized" technology regime (superscript G) will then be

$$\Pi^G \equiv \pi(S, Q) - (w + d)Q,$$

where as above we define $Q^G(S, w + d)$ and $\Pi^G(S, w + d)$ to be the optimal demand and payoffs of the AS sector respectively. Both $Q^G(\ )$ and $\Pi^G(\ )$ increase with S and decrease with $w + d$.

"Marginal" S

At this point, we can define the marginal S, denoted by S^0, to be the size S such that the sector is indifferent between using the localized technology or the GPT. The marginal S is defined implicitly by the following expression

$$\Pi^L(S^0) = \Pi^G(S^0, w + d). \tag{4}$$

We know from the analysis of the previous section that the larger S sectors will prefer localized over generalized choice. Thus the AS using the GPT will be all those with $S \leq S^0$, while all AS sectors with $S > S^0$ will use the localized technology. It is straightforward to see that the marginal S declines with $w + d$. Hence the set of firms that use the GPT increases as d declines or if scale economies in a GPT sector lead to lower w in equilibrium.

10.5.2 General Specialist GPT Sector

The general specialist sector sells the intermediate input Q to the applications sector, once again bearing costs that depend on the total demand. We now add an opportunity for the GS sector to invest in reducing the costs, d, of failures to localize. As the examples in our previous section suggest, these can really be thought of as investments in generalized knowledge, inventions, and technologies. By reducing d, the invention lowers the "economic" distance between all applications sectors and the GPT. In turn this reduces the penalty from using the GPT instead of a more specialized technology.

The GPT sector incurs a fixed cost for reducing d, $C(d, \theta)$, where θ is a parameter indexing the exogenous state of science or engineering. We assume that $C_d < 0$ and $C_{dd} > 0$. The cost of reducing d increases at an increasing rate, and it is natural to think that further generalizations of technologies or knowledge bases are harder to obtain.

The flavor of our analysis continues to be cooperative and cost minimizing, even though we now treat GS and AS as choosing separately. On the revenue side, the GS sector benefits from selling the GPT input Q, at a price w, to all industries or firms that are willing to buy it instead of resorting to their local technology. We assume, consistent with the cost-minimizing structure, that w is set to cover the (average) costs of the GS sector.

10.5.3 Equilibrium: The Game between GPT and AS sectors

To analyze the cost-minimizing solution to this system, we assume that the AS and GPT sector have payoffs consistent with cost minimization and play a static Nash equilibrium game. AS decide whether to use GPT or localized technology depending on their price-taking profit. The GPT sector picks d to maximize its own surplus but takes w as given. Since each AS sector chooses individually, we will focus our results on the size of the subset of AS sectors that uses the GPT good rather than localized technology.

Total demand for the GPT good is the demand by all sectors with $S \leq S^0$. Once again we measure the number of sectors by N. That is, the total number of sectors of type S is equal to $N \cdot f(S)$, where $f(\cdot)$ is the density function introduced above. Then the sales of the GPT sector are

$$Q^{GS}(w+d) = \int_{S=0}^{S^0} NQ^G(S, w+d)f(S)\, dS. \tag{5}$$

The payoffs are

$$\Pi^{GS} \equiv wQ^{GS} - K(Q^{GS}) - C(d, \theta). \tag{6}$$

Now w is taken as given by the GS, but at a level greater than MC (by our average cost pricing rule). This is an incentive to lower d. S^0 is a function of d, and S^0 increases as d declines. Hence the generalist sector's willingness to reduce d arises from the desire to attract new applications' demand to GPT use.

Given the expressions for Π^G and Π^{GS}, we look for the Nash equilibria of the game among the GS sector and the ASs. The strategies of the latter are the level demands for the input Q, whereas the strategy of the former is the level of d.[16] The first-order conditions of the problem are

$$\Pi_Q^G \equiv \pi_Q - (w+d) = 0, \tag{7}$$

$$\Pi_d^{GS} \equiv N \cdot f(S^0) \cdot Q^G(S^0, w+d)S_d^0 \cdot (w - K'(Q^{GS})) - C_d = 0. \tag{8}$$

The first-order condition (7) is the one for all the ASs. The first-order condition (8) is the one for the GPT sector. Note that the latter is evaluated at the marginal S, S^0. Thus the behavior of the generalist-specialist sector is determined by the conditions that characterize the "marginal" application. This last feature comes because our game structure has the GPT taking w and the Q of all the inframarginal ASs as given. We also want to note at this point that our structure is not meant to represent a realistic market game. In fact some readers may be confused by this lack of "realism." But our problem is simply to find the collectively coordinated, cost-minimizing equilibrium, and the concept of Nash equilibrium is a useful one for this purpose. Thus we do not think that this realism is crucial for our argument. At the same time, although we do not put any structure on how the cost-minimizing equilibrium is reached, we do want to note that a cost minimizing industry structure is a strong "attractor."[17]

---

16. To avoid confusion, it is important to recall that the benefits associated with the GPT increase as d decreases, and vice versa.

17. Moreover, if one is very keen about realism in the game, in the lemma below we use the concept of supermodular games developed by Milgrom and Roberts (1990). Milgrom and Roberts (1990) also show that the results of supermodular games can be obtained as the outcome of an "adaptive dynamic" process wherein the strategies of the players are played

Lemma The game between the ASs and the generalist sector is supermodular in (Q, d). Hence there exists a set of Nash equilibria, and these are ranked. The best Nash equilibrium is the one with the largest Q's and the smallest d.

Proof Our game satisfies the conditions for supermodular games in Milgrom and Robert's (1990) theorem 4. First, our strategies are bounded.[18] Second, the cross-partial derivatives of the objectives functions, namely Π^G_{Qd} and Π^{GS}_{dQ}, are both nonpositive. The former, Π^G_{Qd}, is the slope of a demand curve. The latter is nonpositive because, given the average cost pricing rule of the GS sector, and our assumptions about the shape of $K(\ )$, $(w - K'(\ )) > 0$; and from condition (8), the marginal benefit of lowering d is higher the higher is Q because the price cost margin $(w - K'())$ is spread over a larger volume of output—$\Pi^{GS}_{dQ} \leq 0$. Hence, by Milgrom and Robert's theorem 5, the (Q, d) game is supermodular. Moreover $\Pi^G_d = -Q < 0$ (AS likes higher d) and $\Pi^{GS}_Q > 0$ (GS likes higher Q). Using Milgrom and Robert's theorem 7, this implies that the preferred equilibrium by both players is the one with the largest Q and the smallest d. ∎

10.6 Division of Inventive Labor as Limited by the Extent of the Market, II

We now once again see that changes in the total size of the market to which the GPT can be applied ($N \times S$) have opposite implications on the size of the GPT sector according to whether they occur through increases in N or S. The new result here is that the division of inventive labor also shifts distinctly as N and S change. Theorem 10.3 below shows that if N increases, with S constant, the GPT sector expands, in the sense that the optimal d declines and the number of firms using the GPT increases. Theorem 10.4 shows that a proportional increase in the size of all firms (or sectors), S, with N constant, leads to a decline of the GPT sector, with opposite effects on d and on the number of firms using the GPT.

Theorem 10.3 An increase in the number of AS in the economy (N) implies a larger GPT sector, in the sense that (i) d is smaller and (ii) a larger set of firms buys the GPT.

---

over time, and there is "learning" over time. They show that within an adaptive dynamic framework, in any finite strategy supermodular game there exists a date after which the strategies are bounded from above and below by the largest and smallest Nash equilibrium. (See Milgrom and Roberts 1990, thm. 8 and related corollaries.)

18. Both d and Q are bounded if the demand curves for Q as a function of $w + d$ cut both axes.

Proof From conditions (7) and (8) it is easy to see that $\Pi_{QN}^G = 0$ and that $\Pi_{dN}^{GS} < 0$. According to Milgrom and Robert's theorem 6, this says that the set of Nash equilibria increases with N. In our case this means that with greater N, all equilibria, and particularly the best one, entail a larger Q and a smaller d. Also with smaller d, S^0 increases, and therefore more firms buy the GPT. ∎

Thus theorem 10.3 says that increases in the size of the market are a powerful determinant of the rise and expansion of a generalist specialist sector, and correspondingly of generalized knowledge bases and technologies. A more extensive market means a larger number of distinct uses of a general purpose technology. With more distinct uses there is, as we noted in the simpler model, a cost-minimizing incentive for the existence of a GS. We see moreover that the generalist specialist industries have greater incentives to reduce d when their markets are more extensive. The technology process of a diverse market economy involves the creation of GPT industries and systematic attempts to make them more general despite their specialization.

But as theorem 10.4 below suggests that even in the broader framework it is still the case that larger individual uses lead to less not more vertical division of labor. The possibility of making the GPT more general introduced in this section is not sufficient to reverse that result.

Theorem 10.4 A proportional increase in the size of all firms in the economy (S) implies a smaller GPT sector, in the sense that (i) d is higher and (ii) a smaller set of firms buys the GPT.

Proof Transform S into $S^* = \alpha S$, with $\alpha > 0$. The distribution of S^* evaluated at the marginal S^* is $G(S^0) = F(S^0/\alpha)$, where $F(\ )$ is the distribution of S. It is straightforward to see that as α increases the percentage of firms that buy the GPT decreases.

To show that d increases, the density of S^* evaluated at S^0 is $g(S^0) = (1/\alpha)f(S^0/\alpha)$. Replace this expression for $f(S^0)$ in condition (8). The marginal benefit of lowering d decreases with α as long as $(1/\alpha)f(S^0/\alpha)$ decreases with α. This expression decreases with α unless the percentage of firms at the margin, namely $f(S^0/\alpha)$, is declining so rapidly that a reduction in S^0/α implies a significant increase in the number of firms of marginal size.[19] ∎

Theorem 10.4 says that with proportional expansions in the size of all using sectors, there are fewer incentives to utilize generalized technologies.

---

19. Technically the elasticity of $f(\ )$ evaluated at S^0/α must be greater than -1.

As in our simpler model of section 10.2, one can have an analogous result for increasing the size of any particular using sector.

10.6.1 Factors Affecting Expansion of GPT: Advances in Science, θ

Finally we examine the impact of changing the conditions of knowledge representation. We introduced a parameter θ into the cost function for lowering d. We now examine the likely form that θ will take if it is the level of scientific knowledge, and its likely impact on a commercial general specialty.

If anything, advances in science mean greater ability to comprehend a wider set of previously unrelated phenomena within common explanatory frameworks, and this facilitates efforts to reduce the distance among them. As a matter of fact, scientific advances have often created technological linkages among formerly distinct industries. For example, greater understanding of solid state physics during the 1950s led to the development of the transistor, thereby inducing greater commonality among the technological bases of industries such as telecommunications, office equipment, consumer products. Similarly advances in the theory of organic chemistry enabled the German chemical industry during the nineteenth century to link molecular structures to the properties of many different substances. Organic chemistry then became the common scientific and technological basis of sectors such as dyestuffs, pharmaceuticals, and explosives. In a very similar way, after World War II, theoretical advances in polymer chemistry provided the common framework to "design" the molecular structures of new plastics, fibers or rubberlike products.

A related interpretation is that θ measures the size and the quality of the professional bodies that in any given industry or society are dedicated to the production of ideas, and that are specialized in the creation of generalized technologies and knowledge bases. It has been suggested, for instance, that the United States provides a more "scientific" education for software programmers than Japan and that this accounts for some of the difficulties the Japanese software industry faces in producing more basic software templates instead of specific, and often highly customized, applications (Cusumano 1991; Nakahara 1993). Similarly Rosenberg (1997) argues that the systematic training provided by the U.S. universities, MIT in particular, in chemical engineering since the end of World War II has been critical for the diffusion of highly skilled professionals in the field. The training of U.S. chemical engineers involved a solid grasp of the scientific foundations of chemical process design. In turn this created a body of professional expertise that could be employed to design and engineer many different types of chemical plants or refineries.

We can then assume that in the problem of the generalist specialist sector the cost of reducing the distance among the application industries is $C(d, \theta)$, where along with the previous assumptions about C_d and C_{dd}, $C_\theta < 0$, and $C_{d\theta} > 0$. The latter two assumptions characterize the role played by the parameter θ. This parameter reduces the cost of a given d, and it reduces the marginal cost of additional reductions in d. It controls for the "ease" with which the generalist specialist sector can lessen the economic distance of the GPT from its applications. Using the framework of our model, we can state the following theorem:

Theorem 10.5 Advances in science (θ), or any other factor that lowers the marginal cost of reducing d, implies a larger GPT sector, in the sense that (i) d is smaller and (ii) a larger set of firms buys the GPT.

Proof Follows that of theorem 10.3. From the first-order conditions, $\Pi^G_{Q\theta} = 0$ and $\Pi^{GS}_{d\theta} < 0$. Using Milgrom and Robert's theorem 6, the set of Nash equilibria "increases" with θ. Hence, with greater θ, the "best" Nash equilibrium entails a larger Q and a smaller d and thus a larger S^o. ∎

Theorem 10.5 clarifies the importance of factors that increase the ability of the generalist sector to utilize basic knowledge to produce general technologies. Our argument corroborates the conjecture made by Arora and Gambardella (1994). There it was suggested that science is a powerful instrument to codify knowledge in ways that enable industries to link seemingly "distant" products and technologies. This led to a "division of innovative labor" because the fixed cost of producing a given piece of knowledge could be spread over a larger market.

One can also inquire about the comparative statics in d. What if all applications sectors of an economy grow more diverse, in the sense that all of their d's grow? We have examined this question, and it turns out that the answer is ambiguous. The reason for the ambiguity is that there are important offsetting effects. A more diverse (higher d) economy has a higher return to localization in a model like our first one. But it also has a higher return to creation of a science- or engineering-based technology (d-lowering). Either of these can dominate in equilibrium, depending on the size distribution of sectors, the shape of $K(\ )$, and so on.

Finally another natural extension is to the case in which apart from the GPT sector, each individual AS can make investments that help reduce their own d, and these investments are complementary to those made by the GPT sector. From the point of view of our theory, this is a trivial case.

The complementarity assumption about the GPT and AS investments implies that any factor (N, S, or θ) that shifts d in a certain direction would shift the individual d efforts of the AS in the same direction.[20] We wanted to mention this case, however, because in many high-tech industries today users often make such complementary investments. For instance, in the development of new information systems many software companies develop general "templates," and the customization of the template occurs through a "rapid prototyping" process in which the general system is passed on to the buyers who start using it and suggest ways to make it closer to their actual needs.[21] Here d is lowered because of efforts made by both parties. Our analysis suggests that users will be more willing to make complementary investments to customize "generic" products to their own needs in markets where there are many uses (N), high scientific skills (θ), or uses of proportionally smaller scale (S).

10.7 A Modicum of Evidence: $N \neq S$ in Two Interesting GPTs

Our theorems about the relationships between GPTs and the size of markets states that the division of "inventive" labor is associated with markets that feature a greater number of distinct uses of a basic technology. These larger markets expand the boundaries of the GPT by widening its breadth of applications and by encouraging investments that reduce the cost of using the GPT vis-à-vis more specific solutions. In this section we look at two industries that each permit examination of this point from two different perspectives.

First is the software industry. From the founding of the computer industry through the mid 1970s, there was no other notion of a commercial computer than large mainframes. The users were large organizations that could afford the high capital outlay and could justify these costs because of extensive utilization of the machines—such as the U.S. military and large corporations. Software was customized. Consultants, systems integrators, or the employees of a particular user would handcraft software applications for it. The rise of a "packaged" software industry, selling "standard" tools for many users, took place only in the 1970s and especially in the 1980s, and was associated to a good extent with the development of minicomputers and later on of the PC.[22] PCs and mini-

---

20. This is also one of the main results in Bresnahan and Trajtenberg (1995).
21. See, for example, Hofman and Rockart (1994).
22. See OECD (1985) on the development of the software industry. Also see Mowery (1996) for a very interesting international perspective.

Table 10.1
Japan–U.S. Hardware and software comparison (1987)

	Japan	United States
Hardware shipments	21.0	45.6
Large systems	8.7	9.1
Medium systems	3.1	8.7
Small systems	5.0	8.2
Personal computers	4.2	19.6
Software-vendor revenues	13.0	24.8
Total packages[a]	1.4	13.1
Custom software and system integration	10.1	9.6
Facilities management and maintenance	1.4	2.1
Total market	34.1	70.4

Source: Cusumano (1991, p. 49).
Note: In 1987 billion U.S. dollars; $1 = ¥125.
a. These include systems/utilities, application tools, and application packages.

computers, with the implied reduction in the price of computational power, meant a bewildering diffusion of computers among many users employing them for very different purposes. Notably many of these users were small (in the sense of small S), and hence they could not afford the upfront cost that was required to purchase or develop customized software. But there were millions of them, definitely large N.

The time-series flavor of this example is incompletely convincing. There was a good deal of technical advance in computers and in software which may well have shifted out the supply curve of independent software vendors. But consider the following cross-sectional evidence: For many reasons the diffusion of minicomputers and PCs in Japan has been slower than in the United States (Cusumano 1991 or Nakahara 1993). As table 10.1 shows, the comparative value of large hardware systems in Japan vis-à-vis the United States is much bigger than in the case of smaller systems and PCs. The numbers are striking. While the sales of large systems in Japan are approximately as big as in the United States (8.7 vs. 9.1 billion dollars), the U.S. PC market is worth 19.6 billion dollars as opposed to 4.2 billion dollars in Japan. There the computer still appears to be the province of large users, who can afford and manage the large systems, and its diffusion among the vast population of smaller users is slow. At the same time table 10.1 shows that the U.S. market of packaged software dwarfs the corresponding Japanese market: 13.1 versus 1.4 billion dollars. By contrast, the figures about custom software are, again, of comparable size (9.6 vs. 10.1 billion dollars).

One can be skeptical of this evidence given the names of the countries in it. After all, market organization is generally more important in the United States, relationship organization for commerce generally more important in Japan. Another example shows that the inference is not country-specific. The history of the computer numerical control (CNC) machines during the 1980s is similar to that of software, but the positions of the United States and Japan are reversed. CNCs are machine tools whose automated tasks are controlled by a computer. The latter can be easily re-programmed to enable the machine to perform a variety of tasks. CNCs, which first emerged in the early 1980s, advanced the earlier technology of numerical controls (NCs) in which the automatic movements were controlled by computer punch tapes.

The United States pioneered this industry, and CNCs were actually invented in the United States. But while the United States had been the world leader in machine tools since the 1970s, Japanese producers expanded dramatically in the world market during the 1980s and increased considerably their exports into the United States. Particularly, the Japanese were able to enter with smaller, microprocessor-based job-shop CNCs, whereas the U.S. producers had remained with large mainframe- and minicomputer-based CNC machines.

The early NC machines were developed by the U.S. Air Force in the 1950s. Since then their diffusion in the United States occurred largely in two sectors, aerospace and automobile, and within the latter predominately among the Big Three—GM, Ford, and Chrysler. In the 1970s about one-third of the total U.S. market of NCs was in the aerospace sector, and the share of the automobile market was a little smaller (Rand 1994, vol. 1, p. 37). With the introduction of CNCs, the U.S. machine tool producers kept focusing on "large, sophisticated users in the automobile and aerospace industries with the available resources and complex requirements to enable adoption of large, expensive, difficult-to-use mainframe-and minicomputer-based CNC machines" (Rand 1994, vol. 2, p. 116); see also March 1988). This also meant that these machines were largely designed for the special purposes and requirements of these users and that the competencies of many machine tool makers were to a good extent sector-specific (Rand 1994; March 1988).

By contrast, Japanese CNC makers immediately focused on smaller, microprocessor-based CNCs for many more types of users. Fujitsu Automatic Numerical Control (FANUC) rapidly became the world leader, and the firm that set the world standards. What FANUC and other Japanese producers did was to develop machines with fairly standardized, commodity-type characteristics. This enabled them to reach the market of many smaller firms in quite distinct industries that were unable to develop large and expensive customized systems. As a matter of fact, even at the end of the 1980s the number of adoptions of NC or CNC machines by Japanese small–medium-sized firms was about 40 percent higher than of small–medium U.S. firms (Rand 1994, vol. 2, p. 112).

Moreover, because their market was composed of many different buyers with distinct features and needs, Japanese producers made significant investments in "modularizing" their production and design operations. This was a real revolution in the organization of their work and in the design of their products that enabled them to take advantage of economies of scale while still maintaining the ability to customize products to meet customer demands. For instance, they made considerable efforts to identify parts of machines for different uses that could be standardized at little loss in terms of specificity of the application (Rand 1994, vol. 2, p. 13). By mixing and matching standardized components, they could then package machines that were suited for different uses—the classic d-lowering strategy for a GPT.

10.8 Conclusion

We have examined the endogenous emergence of GPTs as the creation of a general specialty where otherwise distinct and localized inputs or bodies of knowledge might have existed. This saves the Smith-Stigler conjecture from being stimulating but false. From a firm, market, and institutional microstructure perspective, the distinction between an extensive market (many customers) and a large individual customer seems quite useful in explaining some important technologies, both historically and in the present. From a long-run growth perspective, the results help explain why the scale economies inherent in GPT are growing more important even as economies get larger. Increased specialization of knowledge has been offset by increased generality in the representation of knowledge. With (nontrivial) commercialization efforts, and with increasingly diverse markets, new GPTs and new general specialties continue to contribute to growth, and growth continues to permit their invention. This counterintuitive but extremely important positive feedback loop would have been seen by Smith as one of God's better accomplishments and by Stigler as one of humankind's.

References

Arora, A., and A. Gambardella. 1994. The changing technology of technical change: General and abstract knowledge and the division of innovative labour. *Research Policy* 23: 523–32.

Arora, A., and A. Gambardella. 1997. Evolution of industry structure in the chemical Industry. In A. Arora, R. Landau, and N. Rosenberg, eds., *Dynamics of Long Run Growth in the Chemical Industry*. New York: Wiley.

Braun, E., and S. Macdonald, eds. 1982. *Revolution in Miniature: The History and Impact of Semiconductor Electronics*, 2d ed. Cambridge: Cambridge University Press.

Bresnahan, T., and R. Gordon, 1997. The economics of new goods: An introduction. In T. Bresnahan and R. Gordon, eds., *The Economics of New Goods*.

Bresnahan, T. and M. Trajtenberg. 1995. General purpose technologies: "Engines of growth"? *Journal of Econometrics* 65: 83–108.

Chandler, A., with T. Hikino, eds. 1990. *Scale and Scope: The Dynamics of Industrial Capitalism*. Cambridge, MA.: Belknap Press.

Cusumano, M., 1991. *Japan's Software Factories: A Challenge to U.S. Management*. New York: Oxford University Press.

Freeman, C. 1968. Chemical process plant: Innovation and the world market. *National Institute Economic Review* 45: 29–51.

Hofman, J. D., and Rockart, J. F. 1994. Application templates: Faster, better, and cheaper systems. *Sloan Management Review* (Fall) 36: 49–60.

Helpman, E., and M. Trajtenberg. 1994. A time to sow and a time to reap: Growth based on general purpose technologies. Tel Aviv Foerder Institute for Economic Research Working Paper 23.

Langlois, R., and P. Robertson, eds. 1995. *Firms, Markets and Economic Change: A Dynamic Theory of Business Institutions*. London: Routledge.

March, A. 1989. The US machine tool industry and its foreign competitors. In M. Dertouzos et al., eds., *Made in America: Regaining the Productive Edge*. MIT Commission on Industrial Productivity, Cambridge: MIT Press.

Milgrom, P., and J. Roberts. 1990. Rationalizability, learning, and equilibrium in games with strategic complementarities. *Econometrica* 58: 1255–77.

Mowery, D., ed. 1996. *The International Computer Software Industry*. Oxford: Oxford University Press.

Nakahara, T. 1993. The industrial organization and information structure of the software industry: A U.S.-Japan comparison. Unpublished manuscript. Center for Economic Policy Research, Stanford University.

Organization for Economic Cooperation and Development. 1985. *Software: An Emerging Industry*. Paris: OECD.

Perry, M. K. 1989. Vertical integration: Determinants and effects. In R. Schmalensee and R. D. Willig, eds., *Handbook of Industrial Organization*. Amsterdam: North Holland.

Finegold, D., ed 1994. *The Decline of the US Machine Tool Industry and Prospects for its Sustainable Recovery*. Santa Monica: Rand Corporation.

Remsberg, C., and H. Higdon. 1994. Ideas for rents: The UOP story, Universal Oil Corp. Des Plaines, IL.

Romer, P. 1986. Increasing returns and long-run growth. *Journal of Political Economy* 94: 1002–37.

Romer, P. 1990. Endogenous technological change. *Journal of Political Economy* 98: S71–102.

Romer, P. 1992. Growth based on increasing returns due to specialization. In K. D. Hoover, ed., *The New Classical Macroeconomics*. Aldershot, UK: Elgar.

Romer, P. Why, indeed, in America? Theory, history, and the orgins of modern economic growth. *American Economic Review* 86: 202–206.

Rosenberg, N. 1963. Technological change in the machine tool industry: 1840–1910. *Journal of Economic History*.

Smith, A. 1776. The Wealth of Nations. London.

Spitz, P. H. 1988. *Petrochemicals: The Rise of an Industry*. New York: Wiley.

Stigler, G. 1951. The division of labor is limited by the extent of the market. *Journal of Political Economy* 59: 185–93.

Von Hippel, E. 1995. "Sticky information" and the locus of problem solving: Implications for innovation. *Management Science* 40: 429–39.

Von Ungern-Sternberg, T. 1988. Monopolistic competition and general purpose products. *Review of Economic Studies* 55: 231–46.

Weitzman, M. 1994. Monopolistic competition with endogenous specialization. *Review of Economic Studies* 61: 45–56.

11

Wages, Skills, and Technology in the United States and Canada*

Kevin M. Murphy, W. Craig Riddell, and Paul M. Romer

In this chapter we examine trends in relative wages in the United States and Canada. In a volume on technology in general, and general purpose technologies in particular, these trends are of interest for two reasons. First, technological change is likely to affect the distribution of income. The data considered here illustrate some of the consequences we can expect from the introduction of new technologies. They also point in the direction of policy options for dealing with them.

Second, labor market data open up a new window on the process of technological change. As the other chapters in this volume suggest, new technologies may arrive at sharply different rates in different decades. There is little doubt that this kind of variation arises at the level of the firm or even at the level of the industry. The interesting conjecture is that it also shows up at the level of the economy as a whole.

The usual way to infer something about variation in the rate of technological change is to use the fact that new technologies raise the first moment of the distribution of income. Variation in the rate of technological change should show up as variation in the rate of growth of average wages, average income, or productivity. However, data on these variables are increasingly suspect because of problems with price mismeasurement that may be getting worse as the economy evolves (e.g., see Griliches 1994). Errors in measures of the price level will bias measures of real output, real wages, or productivity. Changes in the magnitude of these errors will bias measured rates of growth in these variables. At the same time that we try to construct better price measures, we must also

---

We thank Michael Pries, Alan Stark, and Laura Veldkamp for expert research assistance, Larry Katz for helpful conversations, and George Akerlof, Pane Beaudry. Pierre Fortin, Elhanan Helpman, Chad Jones, Richard Lipsey, and Alwyn Young for comments on earlier drafts. The Canadian Institute for Advanced Research supported this work. Riddell also acknowledges financial support from the SSHRC through the Western Research Network on Education and Training.

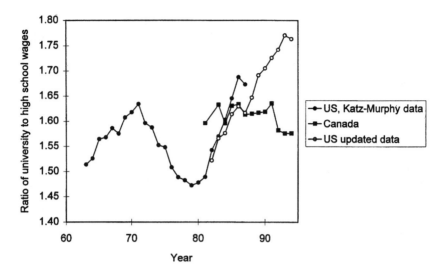

Figure 11.1
Relative wages in the United States and Canada

explore other indicators of the trend in technology that are not susceptible to bias from price level mismeasurement.

If, as many people believe, technological change also influences the second moment of the wage distribution, we can use the trend in relative wages as an independent indicator of underlying changes in technology. As an alternative indicator, the ratio of two wages has the advantage that it does not require a price index and is therefore invulnerable to bias from price mismeasurement. Of course this measure has its own disadvantages. For example, relative wages can tell us about the trend in only one component of technological change—the skill bias. Nevertheless, it is a potentially revealing measure that we should certainly exploit, along with others, when we try to understand the evolution of technology over time.

To get at both of the questions that interest us—How does technology influence relative wages? And, how has the rate of technological change varied over time?—we focus on the university–high school wage premium. We calculate this premium as the ratio of the wage for someone with a university degree to the wage for someone with only a high school education. We illustrate the trends in this premium in figure 11.1. This figure presents data from the United States that were explored by Katz and Murphy (1992). We extend this series with illustrative estimates for the United States that extend through 1994. The figure also presents new data for Canada from 1981 to 1994. (Details of the data we use are presented in section 11.3.) For the rest of the analysis presented here, we work with the Katz-Murphy data through 1987 and process the Canadian data in the same way that that they processed their data. For our purposes, the key fact about the data illustrated in figure 11.1 is that the university wage premium varies substantially both over time and between countries. In the United States it grew during the 1960s and fell during the 1970s. Throughout the 1980s the premium grew in the United States yet fell slightly in Canada.

Many people have interpreted movements in relative wages as an indicator of the forces influencing the demand for different types of labor. In this sense they are implicitly adopting the approach that we advocate here—using movements in relative wages as indicators of changes in relative demands. However, in informal and popular discussions, people typically do not take account of the confounding effects induced by changes in the relative supply of different types of labor. Katz and Murphy have shown that in the United States, the variation in relative wages plotted in figure 11.1 is well explained by changes in the relative supply (1992). Changes in relative supply of university and high school educated workers seem to cause the substantial decade-to-decade variations in the path of relative wages illustrated in figure 11.1. For the 1980s we find similar differences in rates of growth of different types of labor between the United States and Canada. As Richard Freeman and Karen Needels (1993) conjectured, supply changes seem to explain the difference in the behavior of the relative wages of more and less educated workers in these two countries. Our main result is to show that this qualitative impression carries over to a quantitative representation. The same model that Katz and Murphy used to explain the difference in trends over time in the United States also explains the difference between the trend in inequality between the two countries.

After Katz and Murphy correct for relative supply changes, they find no evidence of variation in the rate at which the relative demand for different types of labor has been shifting over time. After we correct for the effects of changes in supply, we find no evidence of differences in the rate at which the relative demand is shifting in Canada and the United States. We also find that over the entire sample period the underlying trend seems to have been increasing relative demand at a steady rate. However, within some subperiods there is some evidence of a slowdown and subsequent speedup relative to the underlying trend. Whether or not this evidence weights against the hypothesis that the process of technological change has varied in a fundamental way in recent decades is discussed later.

Of course it is possible that forces such as increased openness to international trade are shifting the relative demand for labor. It is also possible that the factor bias of technological change (the part of technological change that we measure with relative wages) changes at a steady rate even though the productivity effect of technological change varies from decade to decade. For both of these reasons one must be cautious in interpreting the meaning of the shifts that we infer in the relative demand for labor. The evidence we offer is only one of many different bits of evidence that need to be weighed in making a judgment about overall trends in technological change.

But regardless of how we interpret shifts in the relative demand for labor, it is clear that one cannot infer anything about the magnitude of these shifts, or even their direction, without first correcting for changes in relative supplies of labor. In the data we examine, changes in supply appear to be the most important force causing decade-to-decade and country-to-country variation in trends in relative wages. If this relationship holds more generally, it has important implications for policy responses to growing wage inequality caused by technological change. Policy-makers who want to influence relative wages may find it easier and more productive to adjust the relative supply of workers with different levels of education than the relative demand. If the government adopts policies that raise the educational attainment of some workers, it simultaneously increases the relative supply of more educated workers and lowers the relative supply of less educated workers. Educational subsidies can therefore have a doubly powerful effect on the relative supply. We find a high degree of substitution between different types of workers. Nevertheless, feasible changes in government policies—changes that are of the same magnitude as the differences in policies we see between the United States and Canada—can have a substantial effect on the wage premium, and therefore on income inequality. Moreover other kinds of policies to encourage skill formation are likely to have the same effect.

In the next section we relate our results to work from labor economics, economic history, and growth theory. We also outline some of the main positions in the old political debate about economic prospects for workers in the bottom part of the income distribution. Then we present our basic model and describe our data and results. We conclude with some conjectures about how to reconcile our wage-based finding of no change in the process of technological change with the usual quantity-based finding of a substantial productivity slowdown in the 1970s.

11.1 Technological Change, Income Inequality, and Policy

Although few economists stress the connection, the assumption that new technologies (or new types of capital goods) are relative complements with more educated labor is closely related to the popular intuition that machinery and new technology harm low-skill workers. This concern forms the basis for one side in the long-standing debate about the connection between technology, growth, and income inequality, a debate that goes back at least to Ricardo's discussion of machinery.

In this debate proponents of what one could call the "interventionist" position start from the presumption that technological change harms workers in the bottom part of the income distribution. They conclude that the gains from economic growth will be widely shared only if governments adopt activist policies that resist market forces. Examples of such policies include a minimum wage, regulations limiting work hours, unemployment insurance, workers' compensation, employment standards, and support for unions that negotiate above market wages. Proponents of the interventionist position believe that the advanced industrial economies have been able to achieve a relatively equal sharing of the gains from growth *only* because governments adopted these measures. Many of them fear that a process of global competition will lead to an erosion of these interventions and a corresponding growth in income inequality.

An opposing pole in this debate comes from within the economics profession. Economists do not necessarily oppose the kinds of social welfare policies advocated by the interventionists. Some do, some do not. However, they are close to unanimous that these measures by themselves could not explain much of the gains in standards of living experienced by people in the bottom half of the income distribution. The economists argue that technological change and the market forces it unleashes tend naturally to raise the incomes of all members of the labor force, just as the proverbial rising tide lifts all boats. According to this view, wages for all types of workers will naturally tend to grow as technological change raises the average level of income.

Samuelson's principles textbook gives a representative statement of this rising-tide position as of the mid 1960s:

An alternative theory would ascribe the rise of wages under capitalism to (1) trade-union pressure, (2) government regulation of monopoly, and (3) interventions of a welfare and regulatory kind by democratic governments reacting to militant political pressures from the masses. This cannot be rejected as without substance, for we have seen throughout this book that government actions do have consequences for both good and evil. But the magnitude and pattern of the rise in real wages in this last century has been such as to cast doubt on union or political action as an important element in its explanation.... With the advance of technology and the piling up of a larger stock of capital goods, it would take a veritable miracle of the devil to keep real wages of men from being bid ever higher with each passing decade. Who fails to see this fails to understand the fundamentals of economic history, as it actually happened. (Samuelson 1964, p. 773)

Samuelson is surely correct when he argues that supporters of the interventionist position are wrong about the reasons why income gains have been widely shared since the beginning of the industrial revolution. However, his explanation in terms of capital accumulation and technological change simply fails to confront the microeconomic evidence suggesting that everything else equal, technological progress can harm large parts of the labor force even as it helps others.

The neoclassical vision that Samuelson articulates, one based on the assumption that technology is a complement with a homogeneous labor aggregate, has dominated textbook and policy discussions of wage trends for many years. Economists formalized this view in the neoclassical growth model based on two factors of production: capital and labor. It continues to be embedded in most treatments of aggregate growth. Even growth models that explicitly introduce human capital as a separate factor of production generally do so in a way that makes more educated workers perfect substitutes with less educated workers (Razin 1972; Lucas 1988) or that makes technology a complement with both types of labor (Mankiw, Romer, and Weil 1992; Romer 1990). Either way, there is no scope in such models for technological change to differentially affect wages for workers with different levels of education. In particular, there is no recognition that technological change can reduce wages for some workers.

However, there are two important lines of work in economics that do treat different types of labor as inputs that could be differentially affected by technological progress or capital accumulation. Since the work of Rosen (1968) and Griliches (1969), a number of labor economists have presented evidence supporting the notion that higher-skilled or more educated workers benefit disproportionately from capital investment or technological change. The evidence is sketchy but tends to suggest that more educated workers are stronger complements with new investment than less educated workers. It also suggests that there is a relatively high degree of substitution between these two types of workers. (See Hammermesh 1992 for a summary of research on these questions.)

These conclusions are reinforced by the work of economic historians on education and wages. In the last century the level of educational attainment in the population has increased substantially (e.g., see the discussion of the U.S. experience by Goldin 1994). In the absence of some force that increases the relative demand for educated labor, this increase in the relative supply of more educated workers should have driven their relative wage inexorably downward if more educated workers are not perfect substitutes with less educated workers.

Both the historical evidence (e.g., Katz and Goldin 1995) and the contemporary evidence from labor markets (Hammermesh 1992; Katz and Murphy 1992) suggest that relative wages for more educated workers do fall when the supply of educated workers outpaces growth in demand. This evidence, and everyday experience, both suggest that workers with different levels of education are not perfect substitutes. Four workers with 6 years of education cannot do the job that one worker with 24 years of education can do. Adding more workers with a primary education to the task usually will not help. The elasticity of substitution between more and less educated workers is large, but it is not infinite. They are substitutes, but they are not perfect substitutes.

Nevertheless, the general historical pattern is one with no persistent tendency for relative wages of more educated workers to fall despite the large increases in their relative supply. For more than a century some force must therefore have been increasing the relative demand for more educated workers.

Some economists have pointed to increased international trade as the source of the demand shifts that have taken place in the last couple of decades (Leamer 1996; Wood 1994). Most attempts at quantifying the effects of trade fall far short of explaining the shifts (Sachs and Shatz 1994) or find other evidence that is inconsistent with the trade explanation (Katz and Murphy 1992; Lawrence and Slaughter 1993). Nevertheless, the interpretation of the shifts remains open because it is difficult to offer direct evidence on the effects of technological change (e.g., see Levy and Murnane 1996). The historical evidence is relevant because it reminds us that the pattern of relative demand shifts extends back at least into the last century. Any explanation based on developments in the world trading system since World War II cannot capture the whole story. It also fails to explain why levels of education have been growing throughout the world without driving down the worldwide returns to education.

Technological change is an obvious candidate force that has been impinging on the economy for centuries. As Lipsey, Bekar, and Carlaw suggest in chapter 9, there is no theoretical presumption that technological change will tend to raise the demand for more educated workers.

Nevertheless, the evidence suggests that over long historical periods, it has done so.

In their analysis of historical trends, the proponents of the interventionist interpretation overestimate the power of conventional labor market policies. They also fail to appreciate the overwhelming importance of education as a force that limits the growth in inequality. But in responding to the interventionists, the proponents of the rising-tide view tend to oversimplify, perhaps even overstate, their case. They gloss over both the theory and the evidence, suggesting that technological change can depress both relative and absolute wages for workers in the bottom part of the skill or educational distribution. Like the interventionists, advocates for the rising-tide model also fail to emphasize the importance of increases in education attainment.

In language that they borrow from Tinbergen (1975), Katz and Goldin (1996) suggest that it is more appropriate to think of the behavior of relative wages the as outcome of a race between technological change and increases in educational attainment. The theory and evidence that we present here fits more comfortably with this education-race view than it does with either the interventionist or rising-tide explanations for the interaction between growth and inequality.

In recent years the various positions in the debate about wages and technology have become more complicated than these three positions suggest. The growth in wage inequality in the United States during the 1980s made it increasingly difficult to support a pure version of the rising-tide position on growth and wages. As a result a variety of economists began to consider once again the possibility that technological change could increase inequality. Because they believed that the rising-tide theory fit the historical facts, they suggested that something fundamental about the nature of technology must have changed. This re-examination lent additional importance to the concept of a general purpose technology that was defined by Bresnahan and Trajtenberg (1995) and analyzed in the context of a model of growth by Helpman and Trajtenberg in chapter 3. The introduction of a new general purpose technology is a natural way to explain a sharp change in the nature of technology.

In economic history in general and especially in discussions of technology, there are many antecedents for the concept of a GPT. For example, it is closely related to the concept of "technological convergence" as described by Rosenberg (1963) in his discussion of the central role that machine tools played in the development of industry in the last century. It can also be understood as a refinement and formalization of the concept of a technological trajectory as described by Freeman and Perez (1988).

As it has come to be used, both in this volume and more generally, the concept of a GPT captures two distinct ideas. First, it formalizes the notion that technological change is an irregular process that can have different effects on the economy at different times. Second, the concept of a GPT reminds us that adopting new technologies can be costly and disruptive. In this sense it echoes and partially rehabilitates the old Ricardian and Marxian vision of technology as a double-edged sword.

This first part of the GPT concept—the claim that technological change can be radically different in different periods—offers one possible reason why technological change seemed to have different effects in the 1950s and 1960s as opposed to the 1970s and 1980s. This leads to the fourth and final theoretical position that we identify concerning the relationship between growth and inequality. The "trend break" hypothesis argues that developments in the last two decades represent an important departure from the rest of the post–World War II period. According to the most common statement of this view, the digital computer is the natural candidate for a GPT that began working its way through the economy in the last two decades. Some economists have argued that this approach offers a unified explanation for both the productivity slowdown and the increase in wage inequality (Greenwood and Yorukoglu 1996). Others recognize that it is difficult to explain the difference in the timing of these events with a single technology shock. The slowdown in the United States was evident in the data even before the oil shock in the early 1970s. Wage inequality did not begin to increase until the end of the 1970s. As a result other proponents of the trend-break view have used the arrival of the computer to explain just the productivity slowdown (David 1991) or just the developments in the labor market (Krueger 1993).

There are many possible versions of the trend-break position and combinations of this position with the other positions outlined here. For example, the economy could cycle back and forth. In periods when a new GPT is being introduced, productivity growth slows and income inequality could increase. In periods when existing GPTs are being developed, productivity growth could be more rapid and income inequality could shrink. Yet over long periods of time, the average behavior of the economy could look like one characterized by the rising-tide model (chapter 3). Alternatively, the trend break could be imposed on top of a model of an education race, as suggested by Autor, Katz, and Krueger (1996).

In this broader context we can restate our conclusions. Based on the behavior of wages in the United States and Canada, we see no reason to prefer a trend-break model to a pure model of an education race with a constant rate of technological change. The education-race model seems to fit the historical data and our data on developments over the last few decades and between the United States and Canada. In this sense our evidence does not provide any aggregate level support for the first of the two ideas subsumed in the concept of the GPT—the idea that there are sharp breaks in the process of technological change. This does not deny the importance of irregularity or trend breaks in the analysis of technological change at the microeconomic level or in the analysis of specific technologies like digital computing. It merely suggests that their effects may average out if there are enough different technologies following independent trajectories. At the aggregate level the process may resemble most closely a steady process of technological change.

Although a steady rate of technological advance appears to fit best with our Canadian and U.S. evidence over the entire sample period, we find some evidence of a slowdown and speedup phenomenon during relatively brief subperiods of the U.S. experience. As we note later, this evidence appears consistent with some theoretical predictions of the economywide impacts of GPTs.

In developing this position, we are forced to move well outside the framework of a simple neoclassical model in which one can treat labor as a homogeneous commodity or treat all factors of production as complements with technology. We allow skilled labor and unskilled labor to be substitutes for each other and take seriously the possibility that technological change can reduce the relative or absolute wage for low-skilled workers. It has long been clear that it is theoretically possible for Ricardo's fears to be realized (e.g., see Samuelson 1988). The evidence from the labor market that we examine supports Ricardo's concerns about machinery, or more precisely, it supports a closely related concern about technology. It suggests that this concern is of real significance in practice, not just in principle.

In this sense we support and extend the revisionist interpretation of technological change that is implicit in the concept of a GPT. In the neoclassical growth model and simple endogenous growth models, technology tends to raise the marginal productivity of all factors of production, and growth proceeds in a smooth fashion by scaling up the marginal productivity of all factors of production. The new interpretation implicit in the concept of a GPT is that new technologies cause large changes in the relative marginal productivity of different inputs. These changes in relative returns induce large changes in prices and offsetting changes in patterns of investment throughout the economy. Our interpretation of the evidence in terms of an education race is fully consistent with this second thread in the concept of a general purpose technology.

11.2 Model

The basic model that we use to motivate our analysis starts with a constant elasticity of substitution aggregator that depends on university educated labor, H, and high school educated labor, L. Because we are interested in the effects of technological change, not its sources, we will represent technological change in terms of exogenous functions of time. Specifically we characterize technological change in terms of two functions of time, $A(t)$ and $B(t)$. The first augments the services of university educated workers; the second augments high school-educated workers. Then we nest the expression for labor inside a Cobb-Douglas function that depends on this labor aggregate and physical capital, K:

$$Y = F(K, H, L, t)$$
$$= K^\alpha (\gamma (A(t)H)^{(\sigma-1)/\sigma} + (1-\gamma)(B(t)L)^{(\sigma-1)/\sigma})^{(1-\alpha)\sigma/(\sigma-1)}. \qquad (1)$$

When we use this kind of single-sector, aggregate production function, we must be clear about what it means. At time t, $F(K, H, L, t)$ represents the maximum possible output for an economy with total resources K, H, and L. This kind of aggregate production function is always well defined. However, there is no reason to believe that actual economies satisfy the assumptions from aggregation theory necessary for us to be able to give this aggregate production any simple interpretation in terms of the production functions of individual firms.

For example, the parameter σ, which controls the elasticity of substitution between H and L in the aggregate economy, may not be closely related to the elasticity of substitution faced by an individual firm. Shifts between firms or industries may be more important than shifts within firms. Similarly one cannot necessarily interpret the aggregate augmentation factors $A(t)$ and $B(t)$ in terms of standard microeconomic processes of technological change. At the firm level, one would typically impose the restriction that technological factors that augment inputs like H and L are monotonically nondecreasing. In general, this kind of microeconomic assumption does not imply anything about the behavior of the aggregate functions $A(t)$ and $B(t)$.

We make one final observation. When we apply this model to data from the last four decades, we will interpret more educated workers H as workers with a university degree and less-educated workers L as workers with a high school education. However, if we wanted to apply the model to earlier historical periods, we would bring in additional levels of education. What constitutes a less-skilled worker changes over time. For example, if we let M denote workers with only a primary school education, we could write an extended production function $\hat{F}$ as follows:

$$\hat{F}(K, H, L, M, t) = K^\alpha (\gamma(A(t)H)^{(\sigma-1)/\sigma} + \varphi(B(t)L)^{(\sigma-1)/\sigma} \\ + (1 - \gamma - \varphi)(D(t)M)^{(\sigma-1)/\sigma})^{(1-\alpha)\sigma/(\sigma-1)}.$$

In the decades from the 1960s to the 1990s, a period of time when university graduates became a significant fraction of the labor force, changes in the economy made the factor $A(t)$ grow relatively rapidly and made the factor $B(t)$ grow slowly or shrink. The number of primary educated workers M was sufficiently small that the dominant group in the "less educated" workers is those with high school education. However, at the turn of the century an insignificant fraction of the labor force was university educated. Relatively few workers were high school educated, and they were relatively more educated compared to workers with only a primary school education. At that time changes in the economy were leading to a greater demand for high school workers. In terms of the model, $B(t)$ was growing rapidly compared to $A(t)$ and $D(t)$. (See Goldin 1994, and Goldin and Katz 1995, 1996, for a description of the changes that were taking place.) Thus whether a high school-educated worker is a relatively high-skilled or a relative low-skilled worker depends very much on the epoch one considers.

If we neglect the primary educated workers and return to our basic equation (1), we can solve for the ratio of the marginal products of the two types of labor:

$$\frac{W_H}{W_L} = C \left(\frac{A(t)}{B(t)}\right)^{(\sigma-1)/\sigma} \left(\frac{H}{L}\right)^{-1/\sigma}, \tag{2}$$

where C is a constant. We can define growth rate functions $g_A(t)$ and $g_B(t)$ by the rules

$$g_A(t) = \ln(A(t)), \quad g_B(t) = \ln(B(t)).$$

Then we can take logarithms in equation (2) and write

$$\ln\left(\frac{W_H}{W_L}(t)\right) = c + \frac{\sigma-1}{\sigma}(g_A(t) - g_B(t)) - \frac{1}{\sigma}\ln\left(\frac{H}{L}\right). \quad (3)$$

This is the equation that we will confront with the data. We will be interested in the hypothesis that the data are consistent with a model in which the difference $g(t) = g_A(t) - g_B(t)$ grows linearly with time. If they are, we can write equation (3) as

$$\ln\left(\frac{W_H}{W_L}(t)\right) = c + \frac{\sigma-1}{\sigma}gt - \frac{1}{\sigma}\ln\left(\frac{H}{L}\right) \quad (4)$$

for some constant g. If this restriction holds, we can also write this relation in difference form:

$$\Delta\%\left(\frac{W_H}{W_L}(t)\right) = \frac{\sigma-1}{\sigma}g - \frac{1}{\sigma}\Delta\%\left(\frac{H}{L}(t)\right), \quad (5)$$

where $\Delta\%X$ stands for the percentage change in X.

11.3 Data

11.3.1 United States

Except for the illustrative data used in figure 11.1, the data for the United States are the same as those used in Katz and Murphy (1992), so our description will be relatively brief. We describe the details primarily to illustrate the steps we will take with the Canadian data. Raw data from the United States come from the *March Current Population Survey* (March CPS) covering data for years from 1963 to 1987. Katz and Murphy used two different samples to construct the wage and labor supply series. The preferred wage series tracks as closely as possible the wage for a homogenous group of workers, so the sample is restricted to people with strong labor market attachment. For labor supply they use a much broader sample that reflects total hours, by education category, actually worked in the economy in each year. The labor supply series include hours worked for essentially all people in the March CPS.

The wage series for high school graduates is based on wages for people with 12 years of schooling. The series for university graduates is based on wages for people who report having 16 years of schooling or more. After eliminating people who worked less than 39 weeks in the year prior to the March survey, Katz and Murphy classified workers by sex and years of experience. In each of these cells, they calculated an average wage for each year, then formed a wage for all high school graduates (or all university graduates) by taking a weighted average of wages in the different cells. The weights were based on hours worked by all individuals in the cell, averaged over all years in the sample. Because the weights do not change over time, both the high school and university wage series give an estimate of the average wage for a pool of workers with fixed demographic characteristics.

To construct the labor supply series, they count hours worked for all workers and classify them into four categories based on educational attainment: less than 12 years, 12 years, 13 to 15 years, and 16 or more years of schooling. Labor supply in each of these four categories is calculated by assuming that workers within these categories are perfect substitutes and summing up total hours worked using efficiency units based on average wages for each sex and experience cell.

In the final stage they aggregated hours worked for these four educational categories into two series corresponding to hours of work by high school-equivalent workers and hours worked by university-equivalent workers. Because we do not want to assume that workers with different levels of education are perfect substitutes, we have to be careful not to use the wrong procedure to do this aggregation. For example, if we followed an efficiency units approach, we could assume that a worker with 14 years of education is a more efficient provider of the kind of labor provided by high school graduates, and use the ratio of wages for these workers to high school educated workers to find the scaling factor. Or we could assume that this worker is a less efficient provider of the kind of labor provided by university graduates, and use the ratio of their wages to wages for university graduates to calculate efficiency units for them.

To establish the basic units of measurement, assume that in the United States, individuals with 12 years of education supply, on average, one hour of high school-equivalent labor per hour worked. Similarly assume that someone with 16 or more years of education supplies one hour of university-equivalent labor per hour worked. To deal with the other two categories, Katz and Murphy assumed that someone in educational category i supplied a_i hours, measured in efficiency units, of high school-equivalent labor per hour worked and b_i hours, measured in efficiency units, of university-equivalent labor per hour worked. They estimate the coefficients a_i and b_i by running a regression (with no constant term) of the wage series for people in education category i on wages for high school workers and university educated workers:

$$w_i = a_i w_L + b_i w_H. \tag{6}$$

In this expression, w_L and w_H represent the wage series for the high school and university educated workers, respectively.

11.3.2 *Canada*

There are a number of arbitrary choices one must make in constructing aggregate relative wage and relative supply series. There are valid rationales for different choices and none of these choices seem to make an important difference to the final results. Our goal in processing the data for Canada was simply to follow as closely as possible the treatment that Katz and Murphy used for the data in the United States. However, because the measurement of educational attainment in the Canadian survey differs from that in the United States and because the Canadian data are available for fewer years, the exact procedures we followed differ slightly.

The data for Canada come from the *Survey of Consumer Finances* (SCF), an annual supplement to the *Labour Force Survey* carried out in April of each year. This annual supplement is very similar to the March CPS. Indeed, these two surveys have been used in other comparisons of the Canadian and U.S. labor markets (e.g., see the articles in Card and Freeman 1993). The SCF has data on all individuals 15 years of age and older starting only in 1981. Before that year, hours worked are publicly available only for heads-of-households, yet reported income is for the entire household. In addition this survey did not collect data for 1983. Thus our sample includes the 13 years 1981–82 and 1984–94.

The SCF does not report years of education. Instead it reports five categories of educational attainment: (1) attended primary school, (2) some high school, (3) some postsecondary, (4) diploma or certificate from a community college, and (5) a university degree. Throughout this chapter we follow the Canadian usage and distinguish between a university degree and a college diploma or certificate. A college diploma or certificate typically requires two years whereas a university degree generally requires three or four years. (Because students in the province of Ontario typically continue in high school through grade 13, they may complete the course of study in what one would consider a four-year institution in only three years.) We treat the average worker with *some* high school as the basic unit for high school equivalent labor in Canada. Note that this is a different unit from the one used in the United States, 12 years of schooling. We allow for a correction factor that takes account of this and any other stable differences between the United States and Canada. Naturally we treat the average worker with a university degree as the unit for measuring hours worked by university educated workers.

As for the data from the United States, we use the survey information on earnings and weeks of work during the previous year rather than the month of the survey. Because the SCF does not ask about usual hours worked in the previous year, we simply use weeks worked instead of total hours as our labor supply measure. In the United States the CPS does ask this question, so labor supply for the United States is measured in hours worked. For Canada we discarded any records with nonpositive earnings or weeks worked. We formed an estimate of total labor supply using a sample that includes everyone with positive weeks worked. Following the Katz-Murphy treatment of the U.S. data, we formed a separate sample to estimate the wages and labor supply. In the wage sample we included full-time, paid workers who worked 39 or more weeks and who earned at least one-half of the full-year, full-time minimum wage earnings during the year.

We formed cells based on the five educational attainment categories described above, eight experience categories, and the two sexes. Because the survey gives educational attainment by broad category rather than by years of education completed, we subtracted the average years associated with a given level of attainment from age to form an estimate of years of experience in the labor market for workers in each educational category.

The SCF made a major change in the questions it asked on education for years 1989 and after. Although the education questions were much improved, the pre-1989 and post-1989 data are not comparable. At present, no parallel survey for estimating independent conversion factors is available. We adjusted the data by fitting a trend and a post-1989 dummy to each time series. We adjusted the post-1989 data by the value of the coefficient on the post-1989 dummy. These adjustments do not make much difference for the real weekly wage series but do matter for the weeks worked series. If more data become available, it would be useful to improve on this crude adjustment procedure.

The final step is to aggregate our five labor supply series to two series of labor supply measured in units of high school equivalents and university equivalents. As noted above, we treated people in the some-high-school category as our reference level for measuring high school equivalents.

Table 11.1
Summary statistics on weeks worked by educational category in Canada (%)

	Primary school	Some high school	Some postsecondary	College graduates	University graduates
Fraction of total weeks worked	7	47	10	18	17
Growth rate of weeks worked	−5.3	−0.1	2.4	5.3	5.2

That is, we assumed that the average Canadian worker in the some-high-school category supplied one hour of Canadian high school-equivalent labor per hour worked. One Canadian high school-equivalent hour will differ from one U.S. high school-equivalent hour because we defined high school-equivalent workers in the United States in terms of people with 12 years of education. If there are any differences in the quality of the Canadian and U.S. educational systems, this will also lead to a difference in the two concepts of hours of high school labor supplied in the two countries. Finally labor supply in Canada is in weeks worked. In the United States it is hours worked.

We make an allowance for these differences when we compare the two countries. We can do this because all of our estimates depend only on the rate of growth of H/L, not on its level. In principle, it is possible that a university education has a different market value in the two countries. Our adjustment allows for this possibility as well, provided the difference in quality does not change over time.

Table 11.1 reports the fraction of total weeks worked that are accounted for by each of the five different education categories in Canada. These fractions are based on averages taken over all 13 years for which we have data in the 14 years from 1981 to 1994. It also reports the rates of growth of total weeks supplied by people in each of these categories. Workers with college diplomas or certificates supply slightly more labor than university graduates, and their labor supply grows as rapidly as the labor supply of university graduates.

Our final challenge is to allocate weeks worked by people in the three remaining educational categories: primary school, some postsecondary schooling, and college graduate. As before, we assume that on average, a worker in category i supplies a fixed number a_i of efficiency units of high school-equivalent hours per week worked and a fixed number b_i of efficiency units of university-equivalent labor per week worked.

We have only 13 annual observations on wages for each of the five educational attainment categories. In addition wages in Canada do not exhibit the kind of sharp movements that we see in the United States in the Katz-Murphy sample. For both of these reasons, any attempt to use a regression to estimate the coefficients in an equation like (6) is unlikely to yield very precise estimates. As a result we tried a variety of different estimates for a_i and b_i and found that our conclusions were not sensitive to the values we used.

Workers who have completed a primary school education are relatively easy to handle. In both countries wages of primary school educated workers are little different from wages for people in some-high-school or 12-year categories. In Canada they earn about 98 percent of the wages for the workers with some high school. In the United States workers with less than 12 years of education earn about 94 percent of the wages earned by workers with 12 years of education. In each country we allocated 95 percent of the labor supply of workers in the lowest skill category to labor supply for high school equivalents and allocated none of their labor to university equivalents:

$$a_{\text{primary}} = \frac{\text{average } W_{\text{primary}}}{\text{average } W_{\text{some high school}}} = 0.95, \quad b_{\text{primary}} = 0.$$

These are very close to the weights of 0.94 and -0.04 used by Katz and Murphy. They used weights of 0.69 and 0.29 to allocate work effort by workers with some postsecondary education to high school equivalents and university equivalents, respectively. Because we could not estimate these weights for the some postsecondary and college diploma categories in Canada, we followed the lead of Autor, Katz, and Krueger (1996). In one of their estimates, they set both of the weights a_i and b_i for workers with education beyond high school equal to 0.5. Like them, we allocated one-half of the weeks worked by workers with some postsecondary education to high school equivalents and one-half to university equivalents. We used these same values for people with a college education.

Figure 11.2 illustrates the small difference that it makes when we change the weights for the United States from the ones estimated by Katz and Murphy to the more arbitrary ones we used in both samples in our analysis. It plots both the logarithm of the actual relative wage and the predictions generated by the two different labor supply series. One series uses the Katz-Murphy weights and replicates their regression:

$$\ln\left(\frac{w_H}{w_L}\right) = \text{constant} - \underset{(0.15)}{0.71} \ln\left(\frac{H}{L}\right) + \underset{(0.077)}{0.033}\ t,$$

$R^2 = 0.52.$

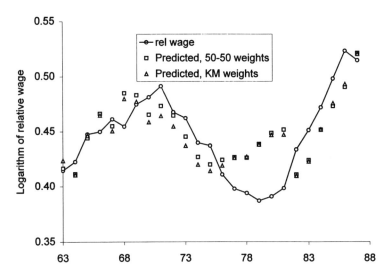

Figure 11.2
Alternative weights for quantity series used with Katz-Murphy data

Standard errors are reported below the coefficient estimates. The relative supply series with the new weights generates slightly different coefficient estimates

$$\ln\left(\frac{w_H}{w_L}\right) = \text{constant} - \underset{(0.16)}{0.81} \ln\left(\frac{H}{L}\right) + \underset{(0.007)}{0.035}\, t,$$

$R^2 = 0.53$.

As the figure shows, these two equations and relative supply series generate very similar predictions for the relative wage. We also checked that our conclusions remain the same if we use the Katz-Murphy weights for both countries instead of the weights described here.

11.4 Estimation and Results

Figure 11.3 illustrates the sense in which we need to impose some restriction on the behavior of $g(t)$ over time to be able to pin down the elasticity σ. If we take equation (3) and solve for $g(t)$, we have

$$g(t) = \frac{\sigma}{\sigma-1}\left\{\ln\left(\frac{w_H}{w_L}(t)\right) + \frac{1}{\sigma}\ln\left(\frac{H}{L}(t)\right) - c\right\}.$$

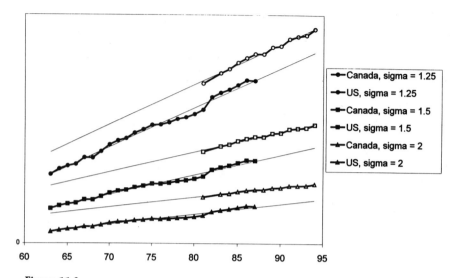

Figure 11.3
Implied values of $g(t)$ and trend line for various values of the elasticity of substitution

Given a value for σ, we can calculate the time path for $g(t)$ up to a constant term, which is of no interest. The figure plots the values of $g(t)$ calculated in this way for different values of σ and for both the United States and Canada. In their analysis of data for the United States, Katz and Murphy found a value of σ equal to 1.41. Thus in this figure we plot data for σ equal to 1.25, 1.5, and 2. We have adjusted the vertical position of the curves to make the graph more readable. The vertical position carries no information. Only the slope matters. In addition to the data series, the figure plots trend lines calculated from a linear regression of $g(t)$ on time.

If, based on a priori information, one specified that $g(t)$ follows precisely the time path associated with one of the values of σ, then inference conditional on this path for $g(t)$ would select precisely this value of σ. In this sense we cannot identify σ without imposing some restriction on the behavior of $g(t)$. There is a one-dimensional continuum of values for σ and paths for $g(t)$, that will all fit the data exactly. Thus, for every possible value of σ, there is a path for $g(t)$ that can rationalize this value. However, the converse is not true. Because the set of possible paths is much larger than the one-dimensional set of possible values for σ, we generally will not be able to find a value for σ that will make an arbitrary path consistent with the data. Thus we can test and reject hypotheses about the behavior of $g(t)$.

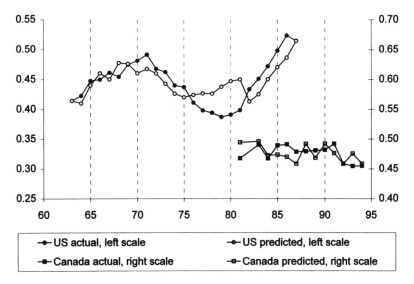

Figure 11.4
Predicted and actual log wages in the United States and Canada

Figure 11.3 also illustrates one of the basic messages from our analysis. In this figure we have not imposed any restriction that forces the slopes of the trend lines to be the same in Canada and the United States for each value of σ. Nevertheless, these slopes are very close. Also, for all of the illustrated values of σ, the path for $g(t)$ fits a straight line surprisingly well. The one exception, to which we will return below, arises when $g(t)$ seems to grow more slowly starting in about 1976 and then grows faster and recovers back to the trend line starting in 1982.

Figure 11.4 shows what happens if we impose the restriction that $g(t)$ is equal to gt and then estimate the parameters σ and g. This figure plots the actual and predicted values of the relative wage in the United States and Canada along with the predicted values from the regression. (To improve readability, note that the values for Canada are plotted against a scale that is translated vertically downward.) The regression estimates behind this equation are

$$\ln\left(\frac{w_H}{w_L}\right) = c_{US} + c_{Canada} - \underset{(0.13)}{0.73} \ln\left(\frac{H}{L}\right) + \underset{(0.006)}{0.032}\, t,$$
$$R^2 = 0.56. \tag{7}$$

We also performed a Chow test to see whether the Canadian and U.S. parameters differ. The F test on the null hypothesis that slope and trend coefficients are the same is $F_{2,33} = 1.2$. If we impose the Katz-Murphy weights on both countries, the corresponding statistic is $F_{2,33} = 1.4$. In either case we cannot reject the hypothesis that the coefficients are the same.

In equation (7) we have allowed for separate constant terms for Canada and the United States. To account for the differences between the data for the two countries, we assume that the true value of H/L in one country, measured in terms of the definitions from the other country, differs from our estimate by a constant factor. When we take logarithms of the relative supply, this factor will be captured in the different values for the constant term. These will pick up any systematic differences between the measured variables in the two countries.

Equation (5), which we repeat below, helps us illustrate why the data impose some restrictions on the trend process even though σ is not, strictly speaking, identified:

$$\Delta\%\left(\frac{W_H}{W_L}(t)\right) = \frac{\sigma - 1}{\sigma} g - \frac{1}{\sigma} \Delta\%\left(\frac{H}{L}(t)\right). \tag{5'}$$

Roughly speaking, we ask whether the data are consistent with the restriction imposed by this equation, and test it against the alternative that the trend rate of growth g takes on different values in the first and second halves of the sample.

Figure 11.5 gives an alternative representation of the data represented in figure 11.2, a representation that is suggested by equation (5). This figure collapses the data into three key periods in the U.S. data and adds a fourth data point for Canada. The three periods in the United States were determined by picking the turning points in the predicted relative wage series in figure 11.4. After aggregating the data into these four points, we can put them on the scatter diagram suggested by equation (5). According to this equation these points should fit on a straight line with slope $-1/\sigma$ and should have an intercept g. As figure 11.2 suggests, the data are roughly consistent with this characterization. We could also fit these four points with four different lines that have the same slope and four different intercepts. These intercepts will correspond to the implied values for g in each period. However, as figure 11.5 shows, we will not be able to do so in a way that makes g consistently different in the first and second halves of the sample. For example, if we make the slope of the line in the figure flatter by making σ larger, it will impose a relatively high value of g

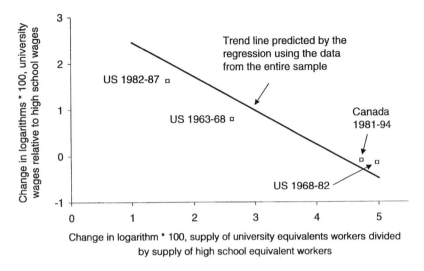

Figure 11.5
Average annual changes in relative quantities and relative wages, measured as log differences ×100

in the 1960s and 1980s in the United States and a relatively low value for g in the 1970s in the United States and in the 1980s in Canada.

If we want to conduct a formal test for a break in the trend, we face the problem of picking the point where we split the sample and the more difficult problem of interpreting the results. Figure 11.6 represents the results from a series of t-tests for a change in the value of g. For each point in the bottom curve in the graph, we estimated a separate regression with a dummy variable interacted with the trend term. In each of these regressions, the dummy allows for a break in a different year. We run these tests for all years from 1968 to 1987. Each point on this bottom curve corresponds to the estimate of this term. It can be interpreted as the change in the value of g for all years after the date corresponding to the point. The points that have a t-statistic that is less than 2 are marked with diamonds and connected with lines. The points with a t-statistic greater than 2 are marked with squares and are not connected with lines.

From this figure we can infer that a test for a trend break starting in 1974 to 1977 would find a statistically significant slowdown in the rate at which the relative demand for more educated labor has been shifting out. In contrast, a test in the years 1981 to 1985 would find a statistically significant increase in the rate at which the relative demand is increasing. This slowdown and speedup are parallel to the slowdown and speedup in

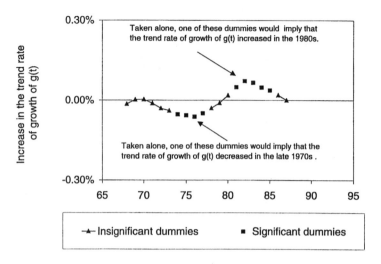

Figure 11.6
Tests for trend break

the paths for $g(t)$ that we observed in figure 11.3. By comparing the predicted and actual values for the real wage, one can see another manifestation of this same process. Between about 1976 and 1981 the changes in relative supplies imply a higher value of the relative wage than the actual value. The estimated trend breaks are another manifestation of the fact that the regression estimates overpredict wages during this period. If we estimate a coefficient for a slowdown in g in 1976 and a speedup in 1981, both of these coefficients are statistically significant. In 1981 the rate returns back to its previous value, as figure 11.3 suggests. This formulation with both a slowdown and a speedup best captures the behavior of the data. Although it is not clear that we should interpret this as evidence of a trend-break, this type of slowdown and speedup phenomenon may be consistent with some theoretical predictions of the impacts of GPTs, such as those in the model developed by Helpman and Trajtenberg (1994).

11.5 Conclusions

There are two ways to interpret the inconsistency between our results and direct productivity measures, which find a significant slowdown after the early 1970s. In the first interpretation, the direct productivity measures are telling us that the rate of growth of the functions $A(t)$ and $B(t)$

has decreased since the 1970s; the labor market evidence is telling us that the rate of growth of the difference $A(t) - B(t)$ does not seem to have changed. That is, productivity growth has slowed down, but the rate of increase in the skill bias in technology has not. In the second interpretation, the rate of growth has not changed for either $A(t)$ or $B(t)$. This explains the labor market evidence which points to no break in the factor bias of technology. The evidence of a productivity slowdown must therefore be the result of price mismeasurement that has been getting steadily worse over time. Our results by themselves offer no basis for choosing between these two explanations. However, until a few years ago the second possibility was given little serious consideration in the economics profession. We interpret our results as one additional bit of evidence suggesting that this second interpretation deserves careful consideration. We need to convert the productivity slowdown from a fact about the economy that needs to be explained into a hypothesis about the economy that needs to be more fully investigated.

Regardless of which of these interpretations is correct, the evidence provided here offers additional support for the claim that over time the cumulative effect of technological change has been to increase the relative demand for more educated workers. The evidence presented here also suggests that the relative earnings of more and less educated workers respond to changes in their relative supply, namely that labor demand is sensitive to the price of labor. Because they can influence the relative supply of more and less educated workers, governments control powerful policies for counteracting this unwanted side effect of technological progress. Policies like those followed in Canada—policies that facilitated during the past two decades substantial growth in postsecondary education at both the college and university levels—can have a major effect on wage inequality. Our results imply that in the absence of this expansion of educational attainment, Canada would have experienced an increase in income inequality between the more and less educated similar to that observed in the United States.

References

Aghion, P., and P. Howitt. 1992. A model of growth through creative destruction. *Econometrica* 60: 323–51.

Autor, D. H., L. F. Katz, and A. B. Krueger. 1997. Computing inequality: Have computers changed the labor market? National Bureau of Economic Research Working Paper 5956.

Bresnahan, T. F., and M. Trajtenberg. 1995. General purpose technologies: "Engines of Growth"? *Journal of Econometrics* 65: 83–108.

David, P. A., et al. 1991. Computer and dynamo: The modern productivity paradox in a not-too-distant mirror. *Technology and Productivity: The Challenge for Economic Policy*. Paris: OECD.

Freeman, C., and C. Perez. 1988. Structural crises of adjustment, business cycles and investment behaviour. In G. Dosi, C. Freeman, R. Nelson, G. Silverberg, and L. Soete, eds., *Technical Change and Economic Theory*. London: Pinter, pp. 38–66

Fortin, N. M., and T. Lemieux. 1997. Institutional change and rising wage inequality: Is there a linkage? *Journal of Economic Perspectives* 11: 75–96.

Goldin, C. 1994. How America graduated from high school: 1910 to 1960. National Bureau of Economic Research Working Paper 4762.

Goldin, C., and L. F. Katz. 1995. The decline of non-competing groups: Changes in the premium to education, 1890 to 1940. National Bureau of Economic Research Working Paper 5202.

Goldin, C., and L. F. Katz. 1996. Technology, skill, and the wage structure: Insights from the past. *American Economic Review* 86: 252–57.

Gottschalk, P. 1997. Inequality, income growth, and mobility: The basic facts. *Journal of Economic Perspectives* 11: 21–40.

Greenwood, J., and M. Yorukoglu. 1997. 1974. *Carnegie-Rochester Conference Series on Public Policy* 46: 49–95.

Griliches, Z. Capital-skill complementarity. *Review of Economics and Statistics* 51: 465–68.

Griliches, Z. 1994. Productivity, R&D and the data constraint. *American Economic Review* 84: 1–23.

Grossman, G., and E. Helpman. 1991. *Innovation and Growth in the World Economy*. Cambridge: MIT Press.

Hammermesh, D. S. 1993. *Labor Demand*. Princeton: Princeton University Press.

Helpman, E., and M. Trajtenberg. 1994. A time to sow and a time to reap: Growth based on general purpose technologies, Centre for Economic Policy Research, Discussion Paper 1080.

Johnson, G. E. 1997. Changes in earnings inequality: The role of demand shifts. *Journal of Economic Perspectives* 11: 41–54.

Katz, L. F., and K. M. Murphy. 1992. Changes in relative wages, 1963–1987: Supply and demand factors. *Quarterly Journal of Economics* 107: 35–78.

Lawrence, R. Z., and M. J. Slaughter, 1993. International trade and American wages in the 1980s: Giant sucking sound or small hiccup? *Brookings Papers on Economic Activity, Microeconomics* 2: 161–210.

Leamer, E. E. 1996. Wage inequality from international competition and technological change: Theory and country experience. *American Economic Review* 86: 309–14.

Levy, F., and R. J. Murnane. 1996. What skills are computers complements with? *American Economic Review* 86: 258–62.

Mankiw, N. G., D. Romer, and D. N. Weil. 1992. A contribution to the empirics of economic growth. *Quarterly Journal of Economics* 107: 407–37.

Razin, A. 1972. Optimum investment in human capital. *The Review of Economic Studies* 23: 455–60.

Ricardo, D. 1963. *The Principles of Political Economy and Taxation.* Homewood, IL: Irwin., chapter 31.

Rosen, S. 1968. Short-run employment variation on class-I railroads in the U.S., 1947–63. *Econometrica* 36: 511–29.

Rosenberg, N. 1963. Technological change in the machine tool industry, 1840–1910. *Journal of Economic History* 23: 414–43. Reprinted in N. Rosenberg, *Perspectives on Technology,* Armonk, NY: M. E. Sharpe, 1976.

Sachs, J., and H. J. Shatz. 1994. Trade and jobs in U.S. manufacturing. *Brookings Papers on Economic Activity, Macroeconomics* 1: 1–84.

Samuelson, P. A. 1964. *Economics: An Introductory Analysis.* New York: McGraw-Hill.

Samuelson, P. 1988. Mathematical vindication of Ricardo on machinery. *Journal of Political Economy* 96: 274–82.

Tinbergen, J. 1975. *Income Differences: Recent Research.* Amsterdam: North Holland.

Topel, R. H. 1997. Factor proportions and relative wages: The supply-side determinants of wage inequality. *Journal of Economic Perspectives* 11: 55–74.

Wood, A. 1994. *North–South Trade, Employment and Inequality.* Oxford: Clarendon Press.

Index

Adoption
 of Internet, 162–64
 late (*see* Laggards)
 of new technologies, 290–91
 order, 106–11
 sectoral wave, 95–96
 strategic complementarities and, 129
Agglomeration economies, 149, 161–62
Application specific sector (AS sector)
 equilibrium with GS sector, 270–72
 model setup, 268–70

Biotechnology, 50
Bronze, historical aspects, 24
Burton process, 182
Business services, 150, 157–58

Canada
 relative wages vs. U.S., 284–85
 technological change model, 292–95
 wage/labor estimation/results, 302–307
 wages/labor supply data, 297–301
Capital
 accumulation, 288
 human, 200, 227, 228
 new, evolution of new GPTs and, 200
Capital obsolescence
 GDP and, 220
 rate, 227–28, 247
Catalytic cracking process, 169, 184, 185–86
Chemical engineering, 167–68, 189–92. *See also* Petrochemical industry
 applications, maturing/expanding, 179–82
 chemical science and, 170–75
 emergence of, 175–78, 264

 expansion, 171
 as GPT, 9, 170, 189–92
 growth, 169
 institutionalization of, 178
 MIT and, 182–87, 190–91, 274–75
 petroleum industry and, 180–82
 training, scientific foundations of, 274–75
 unit operations concept (*see* Unit operations)
Chemical manufacturing
 historical aspects, 265–66
 large-scale, 172–73
Communications
 autarky, 155, 156, 161–62
 face-to-face, 146
 network (*see* Internet)
 semiconductor adoption, 115–16
Complementarities
 chemical engineering and, 190
 gross or Hicksian, 41
 innovational, 3, 16
 net, 41
 strategic, 41, 129
 strong, with existing or potential new technology, 40–43
 technological, 42–43, 214–15
 vertical, 16
Computers
 current ICT revolution and, 23
 diffusion in Japan vs. U.S, 277–78
 knowledge investments, unmeasured, 224–25
 organizational changes for, 73
 for political-military system, 36
 transistor-based, 113–14
Creative destruction, 220

Diffusion
  of minicomputers/PCs, 277–78
  of new GPT, 17–18, 78–79
  order of adoption and, 106–11
  of semiconductor technology, 111–19
  technological change and (see Technological change)
Distillation, 177, 183
Division of labor, 277–79
  cost-minimizing industry structure and, 257–60
  extent of market and, 256, 260–62, 272–76
  general specialties, 253–54, 268–71 (see also General purpose technologies)
  localization and, 255, 268–71
  resulting economies and, 262–67
  stock market and, 254
Dixit-Stiglitz utility function, 17, 214–15
Dubbs process, 265

Education attainment
  in Canada, 297–301
  income inequality and, 289
  increases, 290
  in United States, 295–97
Electricity
  generation, 51
  historical aspects, 27–28, 48, 55
  introduction of, 72–73
Enabling technologies, 3, 20
Endogenous growth model
  numerical version, 238–46
  R&D-based, 229–34, 247
Engineering disciplines, 169. See also Chemical engineering
Experimentation, 133–34, 137–38
Extent of market
  division of labor and, 261–62, 272–76
  increased, 255
  local definition, 260–61
Externalities, 8, 210

Facilitating structure
  elements of, 195, 216
  interaction with, 198–203
  long-term changes, 200–202
  transitional impacts, 202–203
Factor mobility, 147–48, 157
Flexible manufacturing, 46
Freeman-Peraz-Soete theory, 19–20

Generality, vs. localization, 256, 257–60, 267
General purpose technologies (GPTs). See also GPT-based growth model
  appreciative theories, 19–20
  definition of, 15, 16, 32–43, 43, 51–52
  diffusion (see Diffusion)
  features of, 3, 32, 55, 85
  formal theories, 16–19
  in history, 21–30
  identification of, 49–50
  Internet (see Internet)
  lasers as, 44
  near-GPTs, 46–47
  organizational technologies as, 45–46
  origins, 216–17
  technological complementarities and, 214–15
  uncertainty and, 217
General specialist sector (GS sector), equilibrium with AS sector, 270–72
General speciality, 253–54. See also General purpose technologies
GPT-based growth model, 4, 56, 82–83
GPTs. See General purpose technologies
Growth. See also Growth models
  cyclic patterns, 85, 121, 123–27
  driving factors, 1
  endogenous, 207–208
  exogenous, 208–209
  impact of new GPT on, 212, 246–48
  Internet, 146–47
  technological change and, 51
Growth models
  GPT-based (see GPT-based growth model)
  neoclassical, 193–94
  Schumpeterian, 121, 122–23
  structuralist, 193–96
GS sector (general specialist sector), equilibrium with AS sector, 270–72

Hearing aids, transistor-based, 112–13, 118
Helpman-Trajtenberg model (HT model), 121
HT model. See Helpman-Trajtenberg model
Human capital, 200, 227, 228

Industrial concentration, evolution of new GPTs and, 201
Industry structure
  cost-minimizing, 257–60
  with localization, 256

Index   313

Information and communication technologies (ICTs). *See also* Computers
  channels of impact, 199
  computer-based, 20
  current revolution, 23
  historical aspects, 21–23
Innovational complementarities, 3, 16
Innovations
  drastic, 2–3, 4 *(see also* General purpose technologies)
  implementation, growth rate and, 127
  incremental, 2–3
  process, 264–66
  vertical, 122
Integrated circuits, in microprocessors, 266–67
Internal combustion engine, 28
Internet, 165
  adoption, 162–64
  as GPT, 44–45, 145–46
  scale economies in services and, 161–62
  services, pricing for, 155–56

Japan
  CNCs, 278–79
  diffusion of minicomputers/PCs, 277–78
  Fujitsu Automatic Numerical Control (FANUC), 278–79
Just-in-time inventory, 46

Knowledge
  changing conditions, GPT expansion and, 274–76
  depreciation, 223
  national income accounts and, 221–26
Kuznets, Simon, 1

Laggards
  characterization, 117–19
  impediments for, 87
  in semiconductor adoption, 115–19
Lasers, 44
Lateral linkages (spillovers), 110–11
Lewis, W.K., 183, 184–85, 186

Marginal S, 269–70
Mass production, 45
Materials technologies
  channels of impact, 199
  historical aspects, 24–25
Microprocessor, 266–67
MIT, role in chemical engineering, 182–87, 190–91, 274–75

National income
  accounts, knowledge and, 221–26
  actual, 245–46
  measured, 245
  R&D-corrected measured, 245–46
New growth theory, 168, 254
Nuclear power, 50–51
Numerical controls (NCs), 278

Obsolescence. *See also* Capital obsolescence
  calculating, 209–10
  depreciation and, 238
  GPT-induced, 226–29
  in rapid technological change, 235–36
  rate, 241–42
  slowdown size and, 142–44
Obsolescence problem, 223, 225–26, 246
Organic chemistry, 274
Organizational technologies
  channels of impact, 200
  as GPT, 45–46

Penicillin, wartime development of, 187–88
Petrochemical industry, 191
  chemical engineering and, 180–82
  proprietary interests and, 169–70
  rise, 187–89
Petroleum industry. *See* Petrochemical industry
Polymer chemistry, 171–72, 173
Power delivery systems. *See also* specific power delivery systems
  channels of impact, 199–200
  generic functions, 34
  historical aspects, 25–28
Printing, historical aspects, 23–24
Productivity
  growth, 235–36
  output growth and, 228–29
  temporary, Internet adoption and, 163–64

Quality-improvement problem, 223–24, 225, 245

Research & development (R&D)
  coordination of expectations, 106
  costs, 62, 209
  diffusion of GPT and, 86, 93–94
  economic growth and, 248
  endogenous growth model, 229–34
  in GPT-related cycle, 7, 17, 18

Schumpeterian growth model, 121, 122–23

Semiconductors, diffusion, 111–19
Shipping technologies, 211
Skilled workers
  demand for, 153–54
  education, 294
  equality of wages, 160
  mobility and, 147–48, 149
  out-migration, 161
  for R&D, 79
  regional trade and, 150, 151–52
  relative wage rate, 79–82
  secondary innovations and, 6
  slowdowns and, 8, 138–40
  supply, relative wages and, 12–13, 288–89
  technological change and, 292
  virtual mobility and, 147
Slowdowns (slumps), 7–8, 136
  by accelerated capital-obsolescence, 220–21
  delayed or aggregate, 129, 134
  duration, 240, 241, 242
  in GDP growth, 220
  historical evidence, 212
  illusory, 209–11, 247
  in macro productivity growth, 207–12
  magnitude, 137
  of measured GDP growth, 225
  model building approaches and, 213
  national income, 248
  obsolescence and, 226
  size of, 8, 128, 137–44, 240–42, 248
  structural links with GPTs, 219
  timing, 18, 128
  transitional, 206–207
  trend-break view of, 291
Software industry, 274, 276–77
Specialization
  horizontal, 255
  increased, 11, 12
  vertical, 255, 256
Stagnation, economic. *See* Slowdowns
Steam power, 26, 42, 55, 170

Technological change
  accelerated, 219–20
  chemical engineering in, 170–75
  differences, in different periods, 290, 291–92
  education attainment increases and, 290
  GPT diffusion and, 85
  income inequality and, 286–92
  labor market data and, 283

learning and, 167–68
model, 292–95
national income accounts and, 221–26
rapid, 234–246, 248–250
rate variations, 283
relative wages and, 284
revisionist interpretation, 292
in terms of exogenous functions of time, 293
uncertainties and, 48
Technological complementarities, 42–43, 214–15
Technological convergence, 32, 266, 290
Technological progress
  incremental nature, 2
  labor and, 288
  obsolescence and, 227
  trajectories, 240–41
Technologies. *See also* General purpose technologies
  competition among, 215
  definition of, 194–95
  enabling, 3, 20
  facilitating structure, 195
  interactions among, 196–98
  policy structure and, 195–96
  process, 195
  product, 194
  public policy and, 195
  radical, 36–38
Technology-spillover, GPTs and, 129, 130–38
Telecommunications, semiconductor adoption, 115–16, 118
Thermal cracking, 182–83
Transistors, 111–19, 274
Transportation technologies
  channels of impact, 199
  historical aspects, 28–30
Trucking, commercial, 29

Uncertainties
  GPT development and, 48–51, 217
  slowdown mismeasurement and, 210–11
  theories, 48
Unemployment
  frictional, 8
  rate, increased, 128
  time paths, 142, 143
Unit operations, 9, 12, 168, 176–78, 190, 264

Universal Oil Products (UOP), 264–65
Unskilled workers
  manufacturing and, 79
  regional trade and, 150
  relative wage rate, 79–82
  slowdowns and, 8
  supply, in U.S. and Canada, 12
  wages, 292
UOP (Universal Oil Products), 264–65

Vertical disintegration, 253, 255
Vertical integration, 31
  incentives, GPTs and, 10
  large market and, 31, 254
  theories, 255

Wage rates
  diffusion and, 103–106
  introduction of communications GPT and, 148
  mobility consequences, 147–48
  productivity-adjusted, 122
  real, 74, 103–106, 142, 143
  relative, 79–82, 284